Knowledge-based Economies

Władysław Welfe (ed.)

Knowledge-based Economies

Models and Methods

PETER LANG
Frankfurt am Main · Berlin · Bern · Bruxelles · New York · Oxford · Wien

Bibliographic Information published by the Deutsche Nationalbibliothek
The Deutsche Nationalbibliothek lists this publication in the Deutsche Nationalbibliografie; detailed bibliographic data is available in the internet at <http://www.d-nb.de>.

This book was originally published
in Polish under the title: „Gospodarka oparta na wiedzy“.
Copyright by Polskie Wydawnictwo Ekonomiczne,
Łódź/Poland 2007.

ISBN 978-3-631-57981-7

www.peterlang.de

Contents

Preface

The concept of a knowledge-based economy has been introduced to capture major properties of contemporary market economies that apply accumulated knowledge capital on a growing scale. It helps distinguish such economies from the industrial ones that prevailed in the previous centuries. Naturally, industrial economies also used ever-increasing capital of knowledge, but their development was primarily driven by the expansion of fixed capital and the associated improvement of labour force quality. Modelling new, knowledge-based economies is a basic challenge that applied macroeconomics faces today.

This monograph presents an attempt at developing a consistent concept of a macroeconomic model of a knowledge-based economy. It puts emphasis on the discussion of major problems appearing in this area, especially these concerning modelling of aspects such as creation, absorption and use of various forms of knowledge capital. It also tries to sum up the results of the relevant worldwide research, and makes suggestions as to the course of future research, including our own investigations[1].

Chapter 1 introduces a general concept of modelling knowledge-based economies and outlines their desirable structures. Its theoretical underpinning is modern growth theory. Chapter 2 concisely characterizes the neoclassical and endogenous theories of growth, accentuating the role of endogenised technological progress.

Chapter 3 describes the structure and properties of macroeconometric models that have been extended to generate processes in which knowledge capital is created, absorbed and used. The next two chapters present specific questions related to the specification and analysis of the properties of equations that explain the impact of knowledge capital.

Chapter 4 deals with the notion, quantification and generation of total factor productivity (TFP) that reflects the embodied technological progress. Then TFP growth is explained in relation to innovations, and especially vis-à-vis cumulative R&D expenditures, both domestic and transferred from abroad.

Chapter 5 analyses in depth the impact of human capital, and discusses capital measurement issues, as well as efficiency of its use.

The results of the empirical investigation exploring the effects of knowledge capital are mainly based on the international cross-section time series data. The reason for taking this approach was not only a limited length of the available time series, but also the belief that the impacts of applied knowledge capital have a universal meaning. Therefore, the last chapter of the monograph,

[1] The research was developed under the Project of Ministry of Science and Higher Education no. 0232/H02/2005/29.

chapter 6, succinctly characterizes cointegrated models with the cross-section time series data.

Rules underlying the construction of macroeconometric models will be harnessed to reconstruct and extend a long-term model of the Polish economy – W8D. The model will then be used to elaborate the long-term development scenarios for the Polish economy.

The authors wish to thank all persons who contributed their comments on the Polish edition of this monograph, and especially the participants of the Research Seminar on Macromodels at the University of Łódź, as their involvement helped improve the English edition. Of course, all possible errors are the exclusive responsibility of the authors. We also wish to thank the Polish editor (Polskie Wydawnictwo Ekonomiczne) for granting its permission to publish the English version with P. Lang Verlag.

Łódź, March, 2009 Władysław Welfe

Chapter 1

The foundations of modelling a knowledge-based economy

1.1 Introduction

In the last years the economic community has come to a common conclusion that the contemporary market economies gravitate to a structure known as a knowledge-based economy. This concept has been formulated in contrast to an industrial economy system that prevailed in the last centuries (Smith [2002]). Even though the earlier economic systems also took advantage of knowledge that determined their technological progress, at the turn of the 20th century the role of knowledge capital started to dominate in the functioning of economies as a result of automation of manufacturing processes, speedy distribution of management information (recently via the Internet), etc. and in economic growth (endogenisation of technological progress), mainly due to the development of the Information and Communication Technology (ICT).

A major empirical tool for exploring the quantitative properties of the knowledge-based economies and sustainable, long-run socio-economic growth is macromodels, among which macroeconometric models can play a leading role (Garratt et al. [2006]).

Our research in this area started with the construction of a long-term macroeconometric model of Polish economy – W8D (Welfe ed. [2001]) – which was subsequently updated in 2003 (Welfe, ed. [2004]). Because the model has recently been thoroughly restructured and extended, it can be viewed as a new long-term macro-econometric model of the Polish economy W8D-2007 (Welfe, [2008a], [2008b]).

This chapter introduces major properties of the new model, showing then the structure of the satellite models covering the research, technological, educational, ecological, demographic and health sectors. It also presents specifications of the augmented production function, total factor productivity (TFP) equations and human capital as used in the core model. Finally, the pertinent estimation techniques and the likely model applications are presented.

1.2 Macroeconometric models of a knowledge-based economy

The quantitative mechanisms that underlie the growth of a knowledge-based economy can be described empirically by means of adequately expanded macroeconometric models. Such models should draw on economic growth theory which has been enjoying its renaissance, and especially on endogenous and sustainable growth theory. Various aspects of the growth theory have recently been discussed in the Polish literature, mainly by Tokarski [2001]. This area is appealing and hence it attracts theoretical research more frequently (Nahuis [2003]).

Empirical investigations referring to these developments draw on the concept of an empirical model of economic growth developed by Welfe [2000]. This concept gave rise to the building of the long-term macroeconometric model of the Polish economy W8D (see Welfe ed. [2001]), as well as its successive mutations (Welfe, ed. [2004] and Welfe ed. [2008]).

The long-term macroeconometric model of the Polish economy W8D was constructed to encourage studies of the evolution of Polish economy towards a knowledge-based economy and elaboration of long-term scenarios reaching the years 2020-2030. More specifically, the model enabled a thorough analysis of the impacts of endogenous technological progress, or rather of the changes in knowledge capital the progress induces, while allowing for relevant feedbacks. The use of the model and simulation exercises will be presented in separate papers.

The new, long-term model W8D 2007 is a complete structure. The core of its simulation version comprises several blocks of equations, traditionally following the familiar classification of economic activities. The blocks explain:

- final demand, including exports and imports,
- the supply side, including output, and the primary factors of production,
- impact of technological progress (TFP),
- production potential,
- prices and wages,
- financial flows.

The final demand block explains private and government consumption, investments and foreign trade. In the next blocks, the long-run potential output will be generated, depending on fixed capital, labour and TFP determinants, i.e. human capital per employee and cumulative R&D expenditures, both domestic and foreign. The direct and indirect channels of the transfer of foreign knowledge capital – via imports – are distinguished. The indirect channels comprise knowledge capital embodied in imports of investment goods and/or imports of high-tech and low-tech products.

The role of investment as a factor determining an increase in potential output as well as in final demand is emphasized. It will provide the possibility of studying potential disequilibria in the commodity markets in the long run.

The last blocks of the model are prices and financial flows. Prices react to disequilibria and changes in costs, while wages are formed in the course of negotiations. The financial flows are explained within particular institutional sectors.

The economic mechanisms represented in the system of models' equations can be summarized in the following manner. The most important are feedbacks

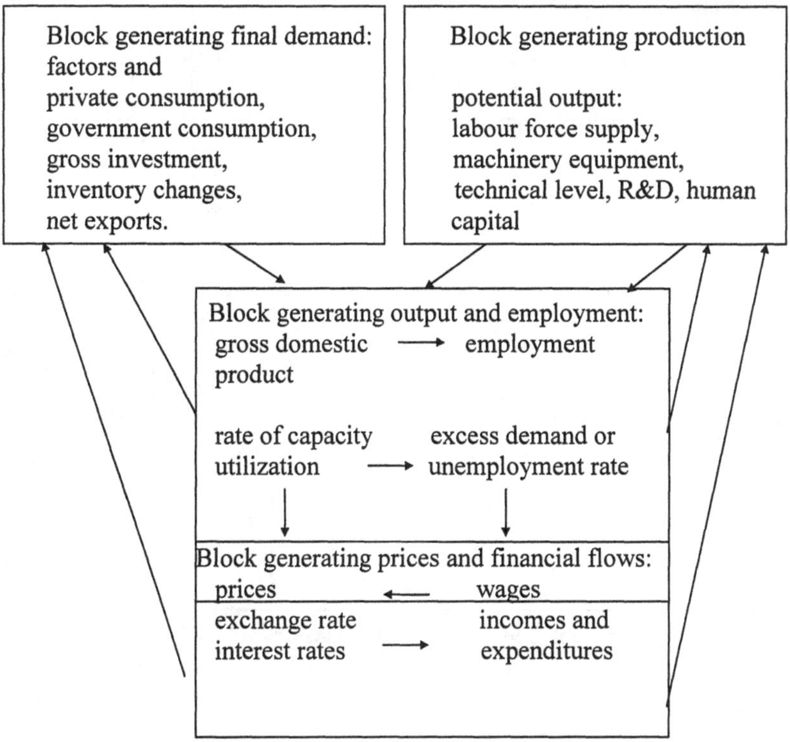

Fig. 1. Structure of model W8

identifying relationships that occur in the production sector between: a) consumption, production and employment (the consumption multiplier), b) investment and production (the accelerator), c) production and the financial sector; the relationships are transmitted, *inter alia*, via the tax system and the budget (the fiscal multiplier), and d) prices and wages (the inflationary spiral). The relationships between the production sector and the financial sector as observed in the early transition economies are mainly taken account of by the quantitative indicators of disequilibria (excess demand, unemployment rates) and in market economies by price and wage adjustments that in turn affect the intensity of the quantitative adjustments (such as changes in demand). Schematic relationships between particular blocks are illustrated in Fig 1.

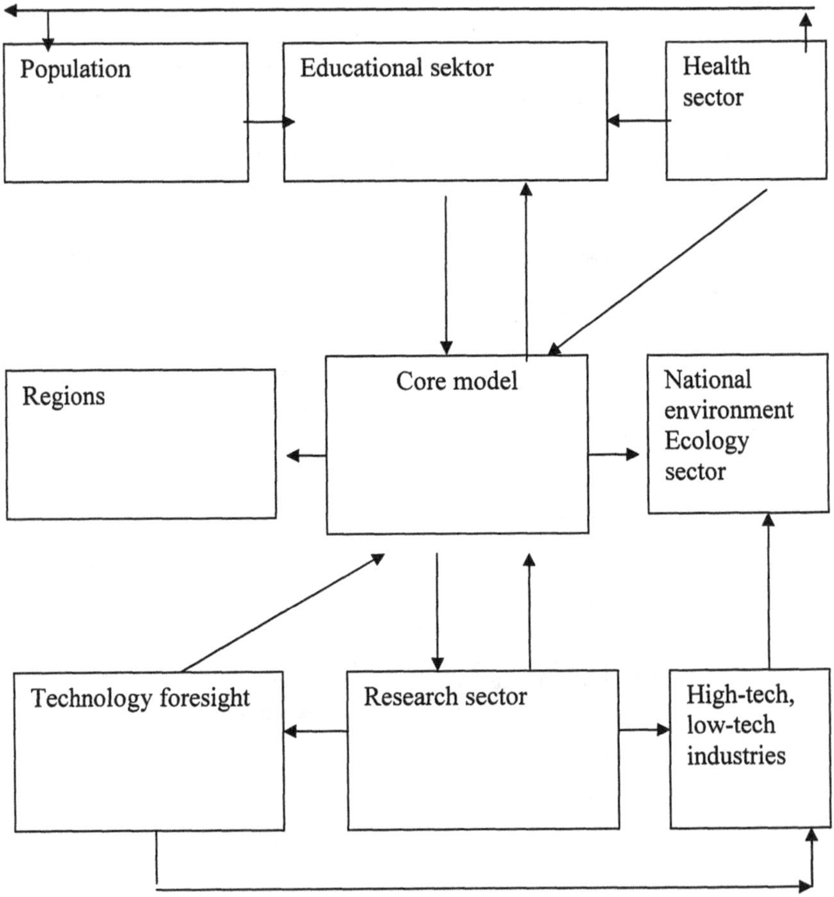

Fig. 2 The core model and the satellite models

When potential GDP is generated in the model, additional relationships are introduced to the real sector. On one hand, the stock of domestic R&D shaped by real domestic expenditures on R&D, and thus by domestic GDP, increases TFP and potential GDP. Likewise, the foreign stock of R&D and additions to foreign R&D transferred with imported machinery influence domestic TFP and GDP. On the other hand, human capital per employee that depends on investment in human capital influences TFP and GDP via increasing real expenditures on education (especially on the tertiary education improving the structure and educational attainment of the graduates). If potential GDP equalled effective GDP, i.e. in the case of a long-run supply driven regime, the model would be provided with a new feedback.

The core model will be extended to include the satellite submodels describing the functioning and growth of sectors such as education, health, and population, as well as research and high-tech industries (with IT), generation of technology (foresight), and ecology (see Fig. 2).

In addition, it is planned to establish links with the regional and sectoral disaggregation of an I-O model, which partly implies application of the cross-section-time series data.

1.3 An extended production function

The main instrument applied to analyze the effects of absorption of knowledge capital is a production function extended to include variables representing changes in knowledge capital.

Following a common practice, we assume that a production process can be described by means of an augmented Cobb Douglas production function with constant returns to scale:

$$X_t^P = BA_t K_t^a N_t^{1-\alpha} e^{\varepsilon_t}, \tag{1.1}$$

where

X^P - potential output (constant prices)

B - constant

A - TFP representing knowledge capital

K - fixed capital (constant prices)

N - employees

a - output elasticity with respect to fixed capital

ε - disturbance term.

When potential output is being explained, it is usually assumed that the factors of production, i.e. fixed capital and employment, are fully utilized. According to this assumption, the effective demand for domestic production X_t is equal to the potential output $X_t = X_t^P$. However, this assumption is only exceptionally met. In reality, potential output is not fully utilized, i.e. $X_t < X_t^P$. Changes in the rate of capacity utilization WX_t usually arise from changes in final demand.

The rate of capacity utilization can be decomposed into the rate of utilization of fixed capital WK_t and the rate of utilization of working time WN_t.

With this information, we can construct a short-term production function explaining effective output. It will have the following form:

$$X_t = BA_t(K_tWK_t)^{\alpha}(N_tWN_t)^{(1-\alpha)}e^{\varepsilon_t}, \tag{1.2}$$

Parameters of the production function (1.2) cannot be estimated without information on the rate of utilization of production factors and total factor productivity A_t. In the early investigations it was assumed that TFP was exogenous and represented by an exponential function of time. Endogenisation of technological progress brings us to the conclusion that TFP dynamics can be explained by making reference to changes in expenditures on the domestic R&D and transferred foreign R&D being the source of new knowledge embodied in fixed capital, and to changes in educational expenditures boosting investment in human capital.

1.4 The dynamics of total factor productivity. Problems of measurement

Following the concept of Solow residual, total factor productivity is commonly used to represent the effects of absorption of knowledge capital and widely applied to international comparisons (Florczak, Welfe, [2000]; Welfe ed. [2001])). However, some measurement problems that constrain its use have not been solved yet (Welfe, [2002], J.Cornwall, W.Cornwall, [2002]).

Firstly, TFP is computed using effective (i.e. observable) output and not potential output. Secondly, several studies, mainly sectoral, use the concept of gross output instead of value added. Further, output elasticities with respect to fixed capital are frequently calibrated and not estimated. All these issues require additional discussions so that empirically acceptable solutions can be reached.

The spectrum of research should be broadened towards sectoral and regional decomposition and it should distinguish between the high and low-tech industries. It may additionally include the multi-sectoral models, such as the input-output ones.

The above questions will be analyzed in some detail in the course of our discussion. Because total factor productivity is not directly observable, Solow residual has been commonly accepted as a starting point for its computation (Solow, [1957]). This residual is derived from the production function where the only explanatory variables are the primary factors of production, i.e. fixed capital and employment. Using the Cobb Douglas production function with constant returns to scale (1.1), we get:

$$X_t^{*} = BK_t^{\alpha}N_t^{(1-\alpha)}e^{\varepsilon_t}, \tag{1.3}$$

where X_t^{*} is potential output obtained under the assumption that technological progress exerts no impact, i.e. $A = 1$.

The rate of TFP growth is assumed to follow the rate of growth of the residual. It is calculated by subtracting the rate of output growth, which was obtained assuming that only the rates of growth of K_t and N_t count, from the rate of growth of output X_t:

$$\overset{\circ}{residual_t} = \left(\overset{\circ}{X}_{tt} - \overset{\circ}{X}_t^{*} \right) = \overset{\circ}{X}_t - \left[a\overset{\circ}{K}_t + (1-\alpha)\overset{\circ}{N}_t \right] \tag{1.4}$$

or in logs:

$$\Delta \ln \quad residual_t = \Delta \ln X_t - [\alpha \Delta \ln K_t + (1-\alpha)\Delta \ln N_t] \tag{1.4'}$$

where X_t is effective output.

Using such defined rate of TFP growth takes us back to the already mentioned, several questions that are solved in different ways (W.Welfe, [2002]). Firstly, the prevailing view is that output is represented at the macro scale by value added (GDP for the national economy). Hence, the only explanatory variables that enter the production function are the primary factors of production – fixed capital and employment. At the micro scale, and often mezzo scale (sectors, branches of industry), the most frequent measures of output are gross output or sales in real terms, mainly because these indicators are more readily available and their calculations are more precise. However, this approach involves an appropriate extension of the production function (1.1). Inputs of energy and intermediate goods must be added as the explanatory variables, i.e. the KLEM production function must be used. Fulfilling this requirement is not an easy task, as information on the use of intermediate inputs is scarce and hardly available for the time series.

Secondly, the implication of using the equation (1.4) is that the value of output elasticity with respect to fixed capital α is known. This value can be arrived at by either estimating parameters of function (1.4) or calibration. In the first case, the function explaining TFP, i.e. A_t, must be specified in detail. In the other case, the results obtained under the neoclassical theory of production (Barro, 1999) are employed; they imply that a value of parameter α can be approximated by the share of surplus in total value added, i.e. by subtracting the share of labour costs from 1. Even though this approach is widely used in empirical research, its application stirs some doubts, because the long-run share of labour costs varies in time and its definitions are not uniform. For instance, different research projects place the share's values for Poland within a range from 0.25 to 0.5 (the latter being close to SNA estimates). This spread of parameters seriously affects evaluation of the rates of TFP growth. For many years the slow rates of growth of fixed capital were associated in Poland with declining employment. After the labour costs were given high weights, the rates of TFP growth showed an upward bias (see W.Welfe, [2002]).

However, the most difficult problem arises when the rates of TFP growth (1.4) are computed with the observed rates of growth of effective GDP (X_t) instead of the unobserved rates of growth of potential GDP (X_t^p). This approach is likely to yield biased estimates of TFP dynamics, because the residual derived from (1.4) reflects not only the impact of technological progress, but also fluctuations in the rate of capacity utilization WX_t. From the definition of WX_t it follows that $X_t = WX_t X_t^p$, hence $\overset{\circ}{X}_t = \overset{\circ}{WX}_t + \overset{\circ}{X}{}_t^p$.

In fact, the rates of TFP growth should be computed using the production function (1.1), which assumes full utilization of the production factors. We then have:

$$\overset{\circ}{A}_t = \overset{\circ}{X}{}_t^p - \left[\alpha \overset{\circ}{K}_t + (1-\alpha)\overset{\circ}{N}_t\right] \tag{1.5}$$

If we transform the expression defining Solow residual (1.4) by replacing $\overset{\circ}{X}_t$ with the term $\overset{\circ}{WX}_t + \overset{\circ}{X}{}_t^p$, we obtain:

$$\overset{\circ}{Residual}_t = \overset{\circ}{WX}_t + \overset{\circ}{X}{}_t^p - \left[a \overset{\circ}{K}_t + (1-\alpha)\overset{\circ}{N}_t\right] = \overset{\circ}{WX}_t + \overset{\circ}{A}_t, \tag{1.6}$$

Therefore, the dynamics of the residual reflects the impact of technological progress only under a constant rate of capacity utilization, i.e. $\overset{\circ}{WX}_t = 0$. This observation is especially meaningful in research investigating countries or periods with changing rates of capacity utilization, mainly following the business cycle. Let us note that the bias could be partly diminished, if we managed to insert some indicators of individual factor utilization, e.g. to swap the number of employees for time worked.

1.5 Specification of the function explaining TFP dynamics. The role of innovations, research and development

The tradition of modelling the TFP growth factors is quite short and many questions still await their answers.

A frequently suggested way of explaining TFP dynamics is decomposition of TFP changes into the effects of readily available capital of knowledge ($\overset{\circ}{A}{}_t^W$), impacts of expanding knowledge capital embodied in fixed capital (A_t^K) and increasing human capital ($\mathring{A}_t^N$). Taking the production function (1.1), we have:

$$\mathring{A}_t = \mathring{A}_t^W + \alpha \mathring{A}_t^K + (1-\alpha) \mathring{A}_t^N \qquad (1.7)$$

The effects of generally available knowledge capital (A_t^W) are either treated as exogenous (usually as an exponential function of time) or attributed to growing knowledge capital associated with improving quality of employment.

The effects of expanding knowledge capital embodied in fixed capital were treated in the past as functions of time; successive generations of fixed capital were assumed to represent rising levels of technology. A simplification was occasionally applied, distinguishing only the share of "new" equipment. The stock of machinery and equipment was frequently used in lieu of total fixed capital. Recently, following the mass computerization of production and management processes, decomposition of fixed capital into computers, computer programs and tele-equipment has been suggested. For instance, this approach was used in studies exploring growth of the US economy (Jorgenson, [2000], Jorgenson, Ho, Stiroh, [2003]), and then in research on the OECD countries (Colecchia, Schreyer, [2002]) and the Netherlands (Leeuwen, Wiel, [2003]). It means, however, that the effects of computerization are removed from the notion of TFP.

When the starting point is decomposition of TFP dynamics as given by (1.7), the impact of TFP embodied in fixed assets (A_t^K) can be related, both directly and indirectly, to the effects of innovations and R&D expenditures. In the first case, patent citations, publications, and technical proximity are used as the approximate explanatory variables of this impact. Viewed in broader terms, however, this impact depends on the anticipated results of R&D expenditures. The cumulative R&D expenditures – both domestic (BRK_t^K) and foreign, transferred from abroad (BRK_t^M) – are assumed to represent the capital of technical and organizational knowledge (Coe, Helpman, [1995]). Further efforts are necessary to answer many specific questions in this area, especially these concerning the transfer of foreign R&D.

The direct and indirect channels through which R&D is transferred are distinguished. Expanding systems of telephone lines, improved technological proximity, increased use of patents, etc., stimulate the direct transfer of R&D (Lee, [2005]), whereas imports of commodities, i.e. of intermediate and investment goods, represents its indirect transfer (Xu, Wang. [1999]). The indirect transfer of knowledge can be summarized by computing the weighted sum of R&D expenditures incurred by the distinguished country *j*:

$$BR_t^M = \sum_j w_j BR_{tj}, \quad (1.8)$$

where w_j represents a weight assigned to R&D expenditures of the country j.

The weights can represent particular countries' shares in the total imports of the analyzed country, or rather the ratios of imports from these countries (i.e. their exports) to their GDPs (Lichtenberg, van Pottelsberghe [1998]). The weights can be linked with imports of intermediate goods (transfer of technology), with imports of investment goods (transfer of new machines, etc.), which seems a better approach (Bayoumi et al., [1999]), or with total imports, when their decomposition is not possible.

In the last 15 years, this research has been given international dimension and its scope includes not only the industrialized countries, but also the developing ones (Engelbrecht, [1997], Bayoumi et al. [1999]). However, the role of FDI in stimulating TFP dynamics continuous to be debatable. It is also necessary to explore further the impact of domestic R&D expenditures and human capital on the rate of absorption of the transferred foreign capital of knowledge. An extension of research programmes to cover TFP development in sectors or branches involves the introduction of intersectoral transfers of knowledge capital.

The above relationships are multiplicative. Hence, the first approximation we can write is:

$$\ln A_t^K = \beta_o \ln BRK_t^K + \beta_1 \gamma \ln BRK_t^M, \quad (1.9)$$

where γ is a weight representing the share of imports, i.e. the degree of openness of the economy.

Because knowledge capital depreciates in time, its value represented by cumulative R&D expenditures will be derived from the balance equation, which in the case of domestic capital reads:

$$BRK_t^K = BRK_{t-11}^K - \delta BRK_{t-1}^K + BR_t^K. \quad (1.10)$$

where

- BR_t^K – current R&D expenditures (constant prices),
- δ – rate of depreciation of knowledge capital embodied in fixed capital that ranges, depending on the author, from 0.05 to 0.15.

It is worth noting that the equation (1.10) provides a link to R&D expenditures financed by the government budget and the enterprise sector.

1.6 Effects of increasing human capital

Investigations conducted in the past to evaluate the impacts of expanding human capital on economic growth produced inconclusive results. The main reason was that they used inconsistent data on the schooling years.

Let us note that many researchers treat human capital as a separate factor of growth and eliminate it from the notion of TFP, which does not seem appropriate.

In general, the scope (i.e. definition) of human capital varies in terms of its coverage. The narrow definition, most frequently used in empirical research, accentuates the differences between the levels of employees' education. The broad version allows for the impacts of learning by doing, health status, etc. In either case, the measurement problems need an adequate solution. Unfortunately, many international projects use simplified measures of human capital, i.e. shares of employees with tertiary education, shares of employees with secondary and tertiary education, or sometimes numbers of students leaving school or simply being educated. This situation has improved in the last years, as more adequate information on the number of schooling years has become available (Fuente, [2004]). Notwithstanding, only few researchers take advantage of the newly developed summary characteristics of human capital per employee.

Such characteristics of human capital (H_t) are designed as the weighted sums of employees with different educational levels i:

$$H_t = \sum_i \mu_i N_{it}, \quad (1.11)$$

where i is the level of education and μ_i is a weight.

The weights may represent:

a) an average number of schooling years, thus H_t stands for total schooling years, or

b) average unit costs of schooling; in this case H_o denotes total educational expenditure,

c) average wages per employee with educational level i; here H_t is an estimate of the total wage bill.

The weights are standardized and the value of the lowest educational level is taken as the base. In most cases, the weights represent the number of schooling years. However, the third approach, where the weights reflect the market efficiency of the level of education, seems to be the most appropriate treatment of human capital functioning as a factor of production (Welfe et. al. [2002]).

Human capital per employee is obtained by dividing total human capital by the total number of employees:

$$h_t = H_t / N_t \quad (1.12)$$

The dynamics of total human capital can be derived from the standard balance equation:

$$H_t = H_{t-1} - \delta H_{t-1} + HJ_t, \quad (1.13)$$

where

HJ_t is investments in human capital (constant prices); the weights represent either educational costs or wages, and δ is the rate of knowledge depreciation.

Seeking relationships between investments in human capital and educational expenditure (BDE_t) is a difficult task. To make this search effective, a submodel describing the educational process and related expenditures has been built (see Welfe et. al. [2002]).

The broad definition of human capital can be applied following an adequate extension of the above measures. The additional characteristics must be related to the central measure. The new characteristics must capture employees' experience (resulting from learning by doing), which is frequently represented by employee age, access to free knowledge (libraries), health status as indicated by life expectancy, etc. (see Benabou, [2002]).

1.7 Estimation and applications of the macroeconometric models

The parameters of the model's equations will be estimated using mainly updated annual time series extending to the year 2005. Most sample data will span the years 1970-2005 and the financial variables will be generally derived from the period 1980-2005. Variables tested for the level of integration are mainly I (1), but in exceptional cases, e.g. for prices, they are I (2).

The estimation of equation parameters may pose several problems that can be dealt with by means of a procedure that basically comprises two steps. First, following Engle and Granger, the long-run parameters will be estimated from the data. Then the ECM procedure will be applied, to allow for the short-term adjustments. The results obtained for both the segments will be analyzed separately. As for several specific relationships, such as TFP, only the short-term equations will be used to explain the residuals.

The simulation version of the model will be used to support the medium and long-term forecasting and scenario analyses extending to the year 2030, or even beyond, until 2040, if the model solutions and the multiplier analysis demonstrate satisfactory stability.

The construction of the new macroeconometric model of a knowledge-based economy will help to launch a new series of long-term forecasts and complementary scenario analyses underpinned by the model-based simulations, thus furthering the earlier simulation exercises (see W. Welfe, ed. [2004]). This new research will take into account the impacts of the long-run increases in domestic R&D expenditure, better absorption of transferred foreign expenditure on R&D, and growth of human capital stimulated by larger financial allocations to the tertiary and post-graduate education.

Consequently, evaluation of the likely effects of realization of particular scenarios drawn up by the administrative authorities will become possible.

References

Barro R.J., 1999. *Notes on Growth Accocunting,* „Journal of Economic Growth", vol. 4, pp. 119-137.

Bayoumi T., Coe D.T.. Helpman E.. 1999, *R&D Spillovers and Global Growth,* „Journal of International Economics„, vol. 47, pp. 399-428.

Benabou R., 2002, *Hitman Capital. Technical Change and the Welfare State,* „Temi di discussione del Servizio Studi", vol. 465, Banca d'Italia, Roma.

Coe D.T., Helpman E., 1995, *International R&D Spillovers,* „European Economic Review", vol. 39, pp. 859-887.

Colecchia A., Schreyer P.. 2002, *ICT Investment and Economic Growth in the 1990s'. Is the United States a Unique Case? A Comparative Study of Nine OECD Countries,* „Review of Economic Dynamics", vol. 3, pp. 408-442.

Cornwall J., Cornwall W., 2002, *A Demand and Supply Analysis of Productivity Growth,* „Structural Change and Economic Dynamics", vol. 13, pp. 203-229.

Engelbrecht H.J., 1997. *International Spillovers. Human Capital and Productivity in OECD Economies. An Empirical Investigation,* „European Economic Review", vol. 41, pp. 1479-1488.

Florczak W., Welfe W., 2000, *Wyznaczanie potencjalnego PKB, łącznej produktywności czynników produkcji* (Potential GDP and TFP), „Gospodarka Narodowa", vol. 11-12, pp. 10-55.

Fuente de la, A., 2004, *Human Capital and Growth: Some Results for the OECD. Current Issues of Economic Growth,* „Proceedings of OeNB Workshops", no. 2, Vienna.

Garratt A., Lee K., Pesaran M.H., Shin Y., 2006, *Global and National Macroeconometric Modelling: A Long-Run Structural Approach*, Oxford University Press, Oxford.

Jorgenson D.W., 2000. *Information Technology and the US Economy,* „American Economic Review", vol. 91. pp. 1-32.

Jorgenson D.W., Ho M.S., Stiroh K.J., 2003. *Lessons for Europe from the U.S. Growth Resurgence,* „CES IFO Economic Studies", vol. 49, pp. 27-47.

Lee G., 2005, *Direct versus Indirect International R&D Spillovers,* „Information Economics and Policy", vol. 17. pp. 334-348.

Leeuwen G. van. Wiel H. van der. 2003, *Spillovers Effects of ICT,* „CBP Report", nr 3, pp. 24-40.

Lichtenberg F., Pottelsberghe de la Potterie van B., 1998, *International R&D Spillovers, A Comment,* „European Economic Review", vol. 42, pp. 1483-1491.

Nahuis R„ 2003, *Knowledge. Inequality and Growth in the New Economy,* E. Elgar, Cheltenham UK.

Smith K., 2002, *What is the 'Knowledge Economy'. Knowledge Intensity and Distributed Knowledge Bases.* "Discussion Paper", series 2002-6, The UN University, INTECH, Maastricht.
Solow R., 1957, *Technical Change and Aggregate Production Function.* „Review of Economics and Statistics", vol. 39, pp. 312-320.
Tokarski T., 2001, *Modele wzrostu endogenicznego* (*Models of endogenous growth*), *in:* W. Welfe (ed.), *Ekonometryczny model wzrostu gospodarczego* (*Econometric Model of Economic Growth*), Wydawnictwo Uniwersytetu Łódzkiego. Łódź.
Welfe W., 2000, *Empiryczne modele wzrostu gospodarczego* (*Empirical Models of Economic Growth*), "Ekonomista", No. 4, pp. 483-499.
Welfe W. (ed.), 2001, *Ekonometryczny model wzrostu gospodarczego* (*Econometric Model of Economic Growth*), Wydawnictwo Uniwersytetu Łódzkiego, Łódź.
Welfe W., 2002, *Łączna produktywność czynników produkcji a postęp techniczny* (*Total Factor Productivity and Technical Progress*), „Studia Ekonomiczne", vol. 36-37, No. 1-2, pp. 94-115.
Welfe W. (ed.), 2004, *Długookresowy, makroekonometryczny model W8D-2002 gospodarki polskiej* (*Long-term Macroeconometric Model W8D-2002 of the Polish Economy*), „Acta UŁ, Folia Oeconomica", vol. 172, Wydawnictwo Uniwersytetu Łódzkiego, Łódź.
Welfe W., 2008a, *The Structure of a New Long-term Macroeconometric Model for Poland,* "Przegląd Statystyczny", vol. 55, No. 4, pp. 5-23.
Welfe W., 2008b, *A Knowledge-Based Economy: New Directions of Macromodelling,* "International Advances in Economic Research", vol. 14, pp. 167-180.
Welfe W. (ed.), 2008, *Makroekonometryczny model gospodarki opartej na wiedzy* (*Macroeconometric Model of Knowledge-based Economy*), "Acta UŁ Folia Oeconomica", Łódź.
Welfe W., Florczak W. Sabanty L., 2002, *Kapitał lodzki i jego endogenizacja* (*Human Capital and its endogenisation*), "Przegląd Statystyczny", vol. 50, No.2, pp. 7-36.
Xu B., Wang J., 1999, *Capital Goods Trade and R&D Spillovers in the OECD,* "Canadian Journal of Economics", vol. 32, No. 5, pp.1258-1274.

Chapter 2

The supply-side determinants of economic growth

2.1 Introduction

Although Ricardo, Marx and Schumpeter are believed to be the precursors of economic growth theory, the first formal models of economic growth were created as late as the 20th century. This group of models comprises one developed by the American mathematician Ramsey in 1928 and another formulated by the Soviet economist Feldman between 1927 and 1928. Examining the contribution of economic growth theory to macroeconomics after Keynesian revolution, Domar ([1962], p. 51) concluded:

"Growth played a strange role in economic theory: it was always around, nearby, but seldom invited to enter the centre. We either treated it as something predetermined, or, quite frequently, it was just an ex-post reflection. For all this time we kept moving forward, discussing employment and investments, interest rate and income, accumulation of capital, economic cycles and many other fascinating problems that obviously require the application of the rate of growth, and for which we tried to find in an extremely ingenious way the readily available solutions in the wonderland of theory".

Economic growth entered modern macroeconomics only when Harrod [1939, 1942], Domar [1962], Kaldor [1971], and Kalecki [1963, 1970] developed their complex analyses on the basis of Keynesian macroeconomics. Their models describe the effect of investment on macroeconomic equilibrium more thoroughly than the traditional model of the Keynesian multiplier or Hicks IS-LM model did, because:

"For Keynes, investment was primarily a demand-driving factor and this was the only way investment affected the level of Gross National Income (...). Theories of growth, however, approach investment more broadly, taking into account also their stimulating influence on the production capacity of an economy. Therefore, theory of economic growth was formulated as a logical consequence of Keynes' theory. When somewhat longer periods are being analysed, the supply-side effect of investment is obvious to discern." (Górski, Sierpiński, [1972], p. 389).

The aforementioned models (as well as Keynesian macroeconomics) were created under a strong influence of the Great Depression in the 1930s, with a view to identifying conditions making the supply-side and demand-side effects of investment counterbalance each other under fully utilized production factors. The major conclusion the Keynesian economic growth models can offer is that a market economy is constantly pressed to abandon a path of growth ensuring macroeconomic equilibrium (this phenomenon is known as a knife-edge problem of the only path of growth where the supply and demand effects of investment balance out). If the government refrains from intervening, a single departure from this path of growth induces a state of permanent, increasing disequilibrium (cf. also Tokarski, [2005a], chapter 1).

As for macroeconomic analysis, the neoclassical models having production functions with decreasing marginal productivities of factors, limited substitution thereof and constant returns to scale followed the Keynesian models of growth. This group of models incorporates, *inter alia*, those developed by Solow [1956], Swan [1956], Phelps [1961], and Mankiw, Romer and Weil [1992]. The development of neoclassical models of growth that Solow initiated was spurred by the critical analyses of models styled after Harrod, Domar, and Kaldor. Having analysed the Keynesian models of growth, Solow ([1988], pp. 307-308) came to the conclusion that particularly Harrod's papers contained many rough statements describing the steady-state growth as a very unstable kind of equilibrium, and that an Earth-bound Martian expedition knowing all the literature on the Keynesian models of economic growth could only expect to see the debris of capitalism that had fallen into pieces long ago. Indeed, the economic history is a register of fluctuations and growth, but most economic cycles seem to have been self-limiting. Hence the conclusion that either the real economies moved along the knife's edge as described in Harrod's or Domar's models, which seems a strange coincidence, or it was the models that fell short of reality they were supposed to describe. Referring to his early works (i.e. from mid-1950s), Solow ([1988], p. 308) wrote about his believing that even if technology itself was not very flexible, the factor-intensity ratios must vary more strongly, because the economy can choose from among the capital-intensive, labour-intensive or land-intensive goods. He also concluded that analysis of the neoclassical production function was more efficient at explaining the processes of long-run economic growth than examination of the long-run adjustment of the demand-side and supply-side of the economy. His model of economic growth additionally demonstrates that in the neoclassical production function economies naturally tend to a steady-state path of growth.

The section below provides a theoretical analysis of factors determining the supply-side factors of economic growth on the grounds of growth models that succeeded the Keynesian models of growth.

2.2 The neoclassical production function and Solow residual

The main instrument used by the Keynesian models of economic growth to explore growth and, implicitly, increasing productivity of labour was a Leontieff production function with constant factor-intensity ratios (i.e. capital and labour intensity). An alternative theoretical approach to analysing the same set of problems builds on production functions where the production processes show varying factor intensity, mainly the so-called neoclassical production functions.

A neoclassical production function $Y=F(K,L)$ is usually understood as a function with the following features:

1. $F(K,0)=F(0,L)$ – this implies that the process of production needs each factor of production (should any factor be missing, then the production process is disabled).
2. $\lim_{K\to+\infty \wedge L>0} F(K,L)=\lim_{L\to+\infty \wedge K>0} F(K,L)=+\infty$ can be interpreted in economic terms that a substantial input (tending to $+\infty$) of one factor and a zero input of the other factor correspond to a substantial stream (tending to $+\infty$) of the product Y.
3. $\partial F/\partial K, \partial F/\partial L>0$ – this means that a growing input of one of the production factors makes, *ceteris paribus*, the product increase. This assumption additionally implies that the marginal products of each factor are positive.
4. Inada's conditions of the form: $\lim_{K\to 0^+ \wedge L>0} \frac{\partial F}{\partial K}=\lim_{L\to 0^+ \wedge K>0} \frac{\partial F}{\partial L}=+\infty$ and $\lim_{K\to+\infty \wedge L>0} \frac{\partial F}{\partial K}=\lim_{L\to+\infty \wedge K>0} \frac{\partial F}{\partial L}=0$ are interpreted as follows: a very small (large) input of each factor corresponds to a very large (small) marginal product of that factor.
5. $\frac{\partial^2 F}{\partial K^2}, \frac{\partial^2 F}{\partial L^2}<0$ – in this case the marginal product of each factor decreases when the factor's input increases. This and the Inada conditions allow concluding that if the input of any of the factors increases from 0 to $+\infty$, then its marginal product decreases from $+\infty$ to 0.
6. The production function is homogenous of degree one. In other words, any ξ-time ($\xi>0$) increase in the output of each factor produces exactly an ξ-time product increase. The economic literature knows this property as constant returns to scale of the production function.

The assumption of constant returns to scale allows us to move from the aggregate production function $Y=F(K,L)$ to the labour productivity function given by $y=f(k)$, where $y\equiv Y/L$ is labour productivity and $k\equiv K/L$ stands for the capital-labour ratio. It can be demonstrated that the labour productivity function $y=f(k)$ has the following properties (see, for instance, Chiang, [1994], pp. 412-414):

1. $f(0)=0$,
2. $\lim_{k\to+\infty} f(k)=+\infty$,
3. $df/dk>0$,
4. $\lim_{k\to 0^+} \frac{df}{dk}=+\infty$ and $\lim_{k\to+\infty} \frac{df}{dk}=0$,

5. $\frac{d^2 f}{dk^2} < 0$.

Owing to the properties of the neoclassical production function, so-called Solow residuals [1957] can be derived that describe the rate of technological progress in the economy (cf. also Harourt, [1975], pp. 97-102). The Solow residuals can be obtained by means of the function *Y=AF(K,L)*, where *A* stands for total factor productivity (TFP), *F(K,L)* is a neoclassical production function and *F*(1;1)=1. Assuming additionally that *Y, A, K* and *L* are continuous and differentiable functions of time $t \in [0;+\infty)$, we can show that[1]:

$$\frac{\dot{Y}}{Y} = \frac{\dot{A}}{A} + \frac{\partial F}{\partial K} \cdot \frac{K}{Y} \cdot \frac{\dot{K}}{K} + \frac{\partial F}{\partial L} \cdot \frac{L}{Y} \cdot \frac{\dot{L}}{L} \tag{2.1}$$

From (2.1) it follows that the rate of growth of the product $\dot{Y}/Y$ is a sum of the rate of technological progress $\dot{A}/A$ (equal to the rate of TFP growth) and the weighted sum of the rates of growth of capital and labour inputs (where the weights are equal to product elasticities with respect to these factors' inputs). If we refer now to Clark's marginal theory (in a perfectly competitive economy every factor of production has to be rewarded according to its marginal product), the factor prices (real interest rate *r* and real wage *w*) will equal $\partial F/\partial K$ and $\partial F/\partial L$, respectively. Inserting the above equations to the dynamic production function (2.1), we arrive at the following equation:

$$\frac{\dot{Y}}{Y} = \frac{\dot{A}}{A} + \frac{rK}{Y} \cdot \frac{\dot{K}}{K} + \frac{wL}{Y} \cdot \frac{\dot{L}}{L} = \frac{\dot{A}}{A} + \alpha_K \frac{\dot{K}}{K} + \alpha_L \frac{\dot{L}}{L} \tag{2.2}$$

where $\alpha_K \equiv rK/Y$ and $\alpha_L \equiv wL/Y$ are the shares of capital and labour in the product.

Because factors' shares in the product have to add up to one, hence, for instance, $\alpha_L = 1 - \alpha_K$. By inserting the above relation to equation (2.2), we have:

$$\frac{\dot{Y}}{Y} = \frac{\dot{A}}{A} + \alpha_K \frac{\dot{K}}{K} + (1-\alpha_K)\frac{\dot{L}}{L} \Rightarrow \frac{\dot{Y}}{Y} - \frac{\dot{L}}{L} = \frac{\dot{A}}{A} + \alpha_K\left(\frac{\dot{K}}{K} - \frac{\dot{L}}{L}\right) \tag{2.3}$$

By substituting $\dot{y}/y = \dot{Y}/Y - \dot{L}/L$ and $\dot{k}/k = \dot{K}/K - \dot{L}/L$ (where $\dot{y}/y$ and $\dot{k}/k$ stand for the rates of growth of labour productivity and capital-labour ratio) in equation (2.3), we can transform and re-write this equation as follows:

---

[1] Notation $\dot{x} \equiv dx/dt$ will henceforth denote a derivative of the variable *x* with respect to time *t*, which on the grounds of economics means an increase in this variable at time *t*. As for the interpretation of the notation $y = f\left(\overset{+}{x_1}, \overset{-}{x_2}\right)$, the variable *y* will be an increasing (decreasing) function of the variable x_1 (x_2), assuming that the other variables will take the fixed values.

$$\frac{\dot{A}}{A}=\frac{\dot{y}}{y}-\alpha_K\frac{\dot{k}}{k} \qquad (2.4)$$

Equation (2.4) is also called the Solow residual. This residual allows estimating the rate of technological progress in the economy. According to the equation, the rate of technological progress $\dot{A}/A$ (or the rate of TFP growth) equals the difference between the rate of growth of labour productivity $\dot{y}/y$ and the rate of growth of capital-labour ratio $\dot{k}/k$ weighted by the share of capital in the product α_K.

In his article written in 1957, Solow calculated the rates of labour productivity growth, the capital-labour ratio and the share of capital in the product (for the non-agricultural private sector in the USA between 1909 and 1949) and estimating that the 1909-1929 rate of technological progress in the USA was around 1% p.a., and 2% in the second half of the analysed period. In other words, around 87.5% of labour productivity growth in the period in question could be attributed to technological progress, while the remaining 12.5% of product growth stemmed from accumulation of physical capital (see Solow, [1957], pp. 312-320; Harcourt, [1975] pp. 97-112).

2.3 The neoclassical models of economic growth

Solow based his neoclassical model of economic growth [1956] on a production function, in opposition to the Keynesian models of economic growth. His model containing Harrodian exogenous technological progress is underpinned by the following assumptions:

1. The process of production is described by the neoclassical production function $Y=F(K,\widetilde{L})$, where $\widetilde{L}\equiv AL$ is a stock of effective labour understood as a product of the number of employees L and the available technological and scientific knowledge A applied to production. Growth in the stock of knowledge immediately augments the productivity of labour. Therefore, this growth represents technological progress as defined by Harrod.
2. The growth of capital stock is the difference between investment (determined by savings) and depreciation of capital. Hence, $\dot{K}=sY-\delta K$, where $s\in(0;1)$ is the rate of saving/investment (i.e. the share of the indicated macroeconomic variables in the product) and $\delta\in(0;1)$ is the rate of capital depreciation.
3. The growth in the stock of knowledge A and the stock of labour L is given by the exogenous rates of growth g and n. Therefore, the stock of effective labour grows at the rate of g+n, i.e. $\dot{\widetilde{L}}/\widetilde{L}=g+n$.

Using these assumptions, we can derive Solow equation given by the formula:

$$\dot{\widetilde{k}}=sf(\widetilde{k})-(\delta+g+n)\widetilde{k} \qquad (2.5)$$

where $\widetilde{k} \equiv K / \widetilde{L}$ is a stock of capital per unit of effective labour and f is a neoclassical function of labour productivity.

The Solow equation can be interpreted that an increase in capital per unit of effective labour $\dot{\widetilde{k}}$ equals the difference between saving/investment per unit of effective labour (i.e. $sf(\widetilde{k})$) and a decrease in capital (that is, $(\delta + g + n)\widetilde{k}$), where the decrease is caused by capital depreciation ($\delta\widetilde{k}$) and growing units of effective labour ($(g+n)\widetilde{k}$). Taking advantage of function f properties and the Solow equation, we can find a stock of capital per effective unit of labour $\widetilde{k}^*$, for which $\dot{\widetilde{k}} = 0$. Moreover, irrespective of the original values of all variables in the Solow model, the economy has to reach the stock of $\widetilde{k}^*$. Therefore, the stock of capital is treated as a stock per unit of effective labour in the Solow long-run equilibrium. From this and the properties of function f it follows that there is also some stream of product per effective unit of labour $\widetilde{y}^*$, to which the economy tends in the long-run.

Because $\widetilde{k} \equiv K / \widetilde{L}$ and $\widetilde{y} \equiv Y / \widetilde{L}$, and assumption 3 provides that $\dot{\widetilde{L}} / \widetilde{L} = g + n$, labour productivity y and capital-labour ratio k grow in the Solow long-run equilibrium at the rate of growth of technological progress as defined by Harrod. It can be additionally shown (see, for instance, Tokarski, [2005a], chapters 1 and 2) that under the conditions of the Solow long-run equilibrium the rates of growth of the main macroeconomic variables are shaped by the rate of exogenous technological progress as defined by Harrods. Secondly, the long-run time paths of labour productivity and capital-labour ratio are positioned the higher, the higher the rate of saving/investment in the Solow model and the lower the rates of capital depreciation and the rates of growth of effective labour units. Hence, a growing rate of saving/investment or falling $\delta+g+n$ in the Solow model may result in a situation, where the rates of growth of the capital-labour ratio and of labour productivity (in a transient period) exceed the rate of growth of Harrodian technological progress. Therefore, in a short period and a medium period the economy ascends to a higher positioned path of economic growth. Having reached it, the rates of growth of y and k run at a level equal to the rate of Harrodian technological progress.

The Solow model of growth inspired Phelps [1961] to create a so-called golden rule for capital accumulation. This rule is understood as a situation in which the Solow model attains a long-run steady-state maximising consumption per employee (assuming that technological progress does not occur in the given economy). In other words, the golden rule for accumulation is given by a rate of saving/investment s_g for which $c^*=(1-s)y^*$ (with c^* and y^* denoting streams of consumption and production per employee under Solow conditions of equilibrium, and s is the rate of saving/investment) reaches its maximum with respect to s. This problem is identical with the maximisation of the relation $c^*=(1-s)f(k^*)$ with respect to k^*. It can be shown that c^* is maximised with respect to k^* when the following relation is true:

$$\frac{df(k^*)}{dk^*} = \delta + n \tag{2.6}$$

According to the above, the golden rule for capital accumulation is a stock of capital per employee k^* for which the marginal product of capital $df(k^*)/dk^*$ equals the depreciation rate of the capital-labour ratio $\delta+n$ (being the same as the rate of depreciation of capital δ plus the growth rate of the number of employees n).

Additionally, the application of the Cobb-Douglas production function to the Solow model leads to a conclusion that the golden rate for saving/investment is $s=\alpha$, where α is the share of capital in the product (the estimates calculated by Solow [1957] and Mankiw, Romer and Wail [1992] show that α is close to 1/3). Therefore, in most real economies the investment/saving rates are considerably below the golden rule for accumulation as formulated by Phelps. This suggests that such economies can boost their long-run consumption per employee by increasing the rates of investment/saving.

Tokarski's works [2003a; 2003b; 2003c] present some attempts at extending the Solow model. In his [2003c] study, Tokarski seeks to identify the long-run paths of economic growth also for increasing returns to scale. According to this study, if the returns to scale are increasing in the economy, then the long-run rate of growth of the basic macroeconomic variables depends not only on the exogenous rate of technological progress, but also on the growth rate of the number of employees (this conclusion holds true also on the grounds of the Mankiw, Romer and Weil model when increasing returns to scale are assumed (cf. Tokarski [2003c; 2005a], chapter 7). This leads to a conclusion that when the returns to scale are expanding, all labour market policies increasing the long-run rates of growth of the number of employees can be an important catalyst of the long-run economic growth.

Tokarski [2003a] attempted to combine the Solow model of growth with the Domar model to find the long-run rules for monetary policy. He assumed in his model that the demand side of the economy was described by an equation similar to the aggregated demand equation that Domar used in his model, and that the supply side followed that used in the Solow model with the Cobb-Douglas production function. He additionally assumed that investment was a decreasing function of a real interest rate.

These Domar-Solow assumptions about economy's functioning can be formally written as Tokarski ([2005a], pp. 124-126):

$$Y^S = K^\alpha \left(e^{gt} L\right)^{1-\alpha}, \quad g > 0, \alpha \in (0;1) \tag{2.7}$$

where:

Y^S – magnitude of potential production
K, L – stocks of capital and labour
α and $1-\alpha$ – the elasticity of Y with respect to K and L
e^{gt} – stock of the available technical and scientific knowledge applied to production
g – rate of technological progress as defined by Harrod.

$$Y^D = m\left(Y^S\right)^\gamma r^{-\beta} \tag{2.8}$$

where:

Y^D – magnitude of aggregated demand in the Domar-Solow economy,
$m>1$ – approximation of the Keynesian multiplier
$\gamma \in (0;1)$ – elasticity of aggregated demand with respect to aggregated supply,
r – real interest rate (as an instrument of monetary policy pursued by the central bank),
$\beta \in (0;1)$ – elasticity of the aggregated demand with respect to the interest rate.

The influence of the interest rate on the magnitude of the aggregated demand in the Domar-Solow model can be explained by the fact that the rate changes affect the magnitude of the economy's aggregated demand (due to consumer and investment demand adjustments, evolving interest rate parity, and the impacts of the Exchange Rate Mechanism transmitted by the net export).

Investment is a decreasing function of the real interest rate. If this assumption is correct, then the equation representing a capital stock increase is given by the following differential equation:

$$\dot{K} = I_0 r^{-\beta} - \delta K \tag{2.9}$$

where $\delta \in (0;1)$ is a rate of capital depreciation.

The Tokarski model ([2003a; 2005a], chapter 6) additionally assumes that the supply of labour (understood as total labour force) grows at the rate of n.

Such designed a model is used to seek a time path for the real interest rates $r(t)$ ensuring the attainment of the following three goals by the economy:

1. The central bank does not allow the aggregated demand Y^D to exceed the

potential product Y^S, because a surplus demand would trigger inflationary pressures.

2. The central bank adjusts the aggregated demand magnitude in the economy to the level of potential production, which counteracts the underutilization of economy's production capacity.

3. Assuming that the rate of unemployment at initial time $t=0$ is the rate of equilibrium unemployment, the central bank does not allow the first rate to change. This implies that the central bank has to find a time path for the real interest rates that will enable the number of employees to grow at the rate of n.

Based on equations (2.7)-(2.9) and the above rules determining the behaviour of the central bank, we can demonstrate that a time path of real interest rates exists, which solves the following equation:

$$G_r = -\frac{(1-\gamma)(1-\alpha)(g+n)}{\beta} - \frac{(1-\gamma)\alpha}{\beta} G_K \tag{2.10}$$

where:

$G_r \equiv \dot{r}/r$ – rate of growth of real interest rates,
$G_K \equiv \dot{K}/K$ – rate of growth of the stock of capital in the Domar-Solow economy.

Equation (2.10) describes the rate of growth of real interest rates ensuring full utilization of production capacity in the economy. This equation provides the following conclusions. Firstly, the central bank should not launch the policy of long-run increases in interest rates until the rate of capital decrease starts exceeding the value of $(1-\alpha)(g+n)/\alpha$. Secondly, a zero rate of capital growth should bring about a reduction in the real interest rates. Thirdly, because equation (2.10) shows that $dG_r/dG_K<0$, the central bank should respond to the long-run growth in capital stock by lowering the real interest rates.

Additionally, equations (2.7)-(2.10) can be used to derive the equation of motion for the rate of capital growth of the discussed model. This equation is a differential Riccati equation of the form Tokarski ([2005a], pp. 128-129):

$$\dot{G}_K = -a\left(G_K - \frac{b}{a}\right)(G_K + \delta) \tag{2.11}$$

where $a=1-\alpha(1-\gamma)>0$ and $b=(1-\alpha)(1-\gamma)(g+n)>0$.

The differential equation (2.11) has a solution according to which the long-run rate of growth of capital stock G_K^* (understood as the rate of capital growth when $t \to +\infty$) stabilises at the level of $G_K^* = \lim_{t \to +\infty} G_K(t) = \frac{(1-\alpha)(1-\gamma)(g+n)}{1-\alpha(1-\gamma)}$ (Tokarski, [2005a], pp. 128-133).

Using Riccati's differential equation (2.11) we can determine, *via* the Cobb-Douglas production function (2.7), the time path of the rate of product growth and the long-run rate of product growth. The long-run rate of product growth obtained from the described model of economic growth is given by: $G_Y^* = \lim_{t \to +\infty} G_Y(t) = \frac{(1-\alpha)(g+n)}{1-\alpha(1-\gamma)}$.

Additionally, on the grounds of the Domar-Solow model both the time path of the rate of growth of real interest rates and the long-run rate of growth of interest rates can be found. The aforementioned long-run rate of growth of real interest rates is as follows: $G_r^* = \lim_{t \to +\infty} G_r(t) = -G_K^* / \beta$.

The model of monetary policy rules in the Domar-Solow type economy seems to offer the following conclusions. Firstly, there is a time path of the long-run interest rates that allows the economy to utilize fully its production capacity. Secondly, under the Domar-Solow equilibrium respecting the above monetary policy rules, the long-run growth rates of capital stock and product stream have some predetermined level (as in the neoclassical growth models by Solow, or in the Mankiw-Romer-Weil model discussed below) that primarily depends on the rate of Harrodian exogenous technological progress and on the growth rate of the number of employees. Thirdly, the solution provided by the Domar-Solow model plots less inclined long-run time paths for capital and product than the Solow or Mankiw-Romer-Weil models do. It is so, because the neoclassical models omit the demand constraints (related to economic growth) and only analyse the paths of potential product. However *"in the reality being described, a theoretical assumption that the observed production corresponds to the potential production is usually untrue"* (Welfe, [2004], p. 64). After adding demand constraints curbing economic growth, the solution exhibits smaller long-run rates of economic growth than in the neoclassical models.

Another option consists in combining the models of growth designed by Solow and Mankiw, Romer, Weil with the Solow [1979] and Summers [1988] models of efficiency wages (cf. Tokarski, [2003b]). The two types of models are put together with a view to endogenising labour demand and the number of employees within a growth model. The major conclusions provided by the procedure are as follows. Firstly, when economic growth shows long-run equilibrium, the rates of growth of the number of employees are positively determined by the rate of physical capital growth (or by the rates of growth of physical and human capital, depending on the specification of the production function), the rate of technological progress and the rate of growth of exogenous labour supply. Secondly, the long-run rate of growth of the number of employees is the higher, the less sensitive wages are to changing labour productivity and the rate of unemployment.

In the 1990s, the leading representatives of new Keynesian economics – Mankiw, Romer and Weil – designed a model drawing directly on the Solow neoclassical model (Mankiw, Romer and Weil called it an augmented Solow model). This model was an extension of the Solow model in the sense that the Cobb-Douglas production function (in the neoclassical model of growth) was added another production factor (human capital) and the model also received an endogenous equation accounting for an increase in the stock of human capital.

In the Mankiw-Romer-Weil model, the following assumptions are made:

1. The magnitude of production in the economy is given by the Cobb-Douglas production function with constant returns to scale (with respect to K, H and $\widetilde{L} \equiv AL$) of the form: $Y = K^{\alpha} H^{\beta} \widetilde{L}^{1-\alpha-\beta}$, where H and K are the magnitudes of human and physical capital in the economy, $\widetilde{L} \equiv AL$ – units of effective labour, and α, β, $(1-\alpha-\beta) \in (0;1)$ stand for the elasticities *of* Y with respect to K, H and $\widetilde{L}$, or for the factors' shares in the product.
2. Increases in the stocks of K and H are given by the differential equations $\dot{K} = s_K Y - \delta_K K$ and $\dot{H} = s_H Y - \delta_H H$, where $s_K \in (0;1)$ and $s_H \in (0;1)$ are the rates of investment in these stocks of capital, and δ_K, $\delta_H \in (0;1)$ are the rates of depreciation.
3. The stock of effective labour $\widetilde{L} \equiv AL$ expands at the rate of $g + n$, where g stands for the rate of exogenous technological progress as defined by Harrod, and n is the rate of growth of the number of employees.

Using $\widetilde{y} \equiv Y/\widetilde{L}$, $\widetilde{k} \equiv K/\widetilde{L}$ and $\widetilde{h} \equiv H/\widetilde{L}$ to denote, respectively, the stream of productivity, the stock of human capital, and physical capital per unit of effective labour, we can write equations representing increases in the stocks of K and H as follows:

$$\left.\begin{aligned} \dot{\widetilde{k}} &= s_K \widetilde{y} - (\delta_K + g + n)\widetilde{k} \\ \dot{\widetilde{h}} &= s_H \widetilde{y} - (\delta_H + g + n)\widetilde{h} \end{aligned}\right\} \qquad (2.12a)$$

The above simultaneous differential equations correspond to the Solow equation (2.5). Furthermore, the investigated simultaneous equations as well as assumptions about the production function in the Mankiw-Romer-Weil model lead to the following simultaneous equations:

$$\left.\begin{aligned}\dot{\tilde{k}} &= s_K\tilde{k}^{\alpha}\tilde{h}^{\beta} - (\delta_K + g + n)\tilde{k} \\ \dot{\tilde{h}} &= s_H\tilde{k}^{\alpha}\tilde{h}^{\beta} - (\delta_H + g + n)\tilde{h}\end{aligned}\right\} \tag{2.12b}$$

The simultaneous differential equations (2.12b) determine the M-R-W model's equations of motion that describe changes in time in the stocks of $\tilde{k}$ and $\tilde{h}$, which correspond to the exogenous rates of investment s_H and s_K , the depreciation rates of human capital δ_H and physical capital δ_K, the rate of technological progress g, and the rate of growth of the number of employees n. Using a phase diagram for the analysed model, we can show that the position of the long-run equilibrium of the simultaneous equations is stable and meets the following relations (Romer, [1996], pp. 130-131):

$$\tilde{h}^* = \tilde{h}^*\left(\overset{+}{s_H}, \overset{+}{s_K}, \overset{-}{\delta_H}, \overset{-}{\delta_K}, \overset{-}{g}, \overset{-}{n}\right) \tag{2.13a}$$

$$\tilde{k}^* = \tilde{k}^*\left(\overset{+}{s_K}, \overset{+}{s_H}, \overset{-}{\delta_K}, \overset{-}{\delta_H}, \overset{-}{g}, \overset{-}{n}\right) \tag{2.13b}$$

where $\tilde{h}^*$ and $\tilde{k}^*$ are stocks of human capital and physical capital per unit of effective labour in the M-R-W long-run equilibrium.

Using relations (2.13a,b) and the production function $Y = K^{\alpha}H^{\beta}\tilde{L}^{1-\alpha-\beta}$, we can demonstrate that a following relation takes place (cf., for instance Tokarski, [2005a], chapter2):

$$\tilde{y}^* = \tilde{y}^*\left(\overset{+}{s_H}, \overset{+}{s_K}, \overset{-}{\delta_H}, \overset{-}{\delta_K}, \overset{-}{g}, \overset{-}{n}\right) \tag{2.13c}$$

Relations (2.13a,b,c) and the fact that $y \equiv Y/L = A\tilde{y}$, $h \equiv H/L = A\tilde{h}$ and $k \equiv K/L = A\tilde{k}$ offer the following conclusions. Firstly, the M-R-W model (like the Solow model analysed before) naturally tends to a steady-state equilibrium, where the stream of labour productivity (y), the stocks of human capital per employee (h) and the capital-labour ratio (k) will grow at a rate corresponding to the rate of growth of Harrodian technological progress. Secondly, the position of the growth path along which the analysed economy will move depends on the rates of investment in stocks K and H, the depreciation rates of these stocks, and the rate of growth of effective labour $g+n$. The higher rates of investment analysed in the Mankiw-Romer-Weil model (i.e. investment in both stocks of capital distinguished in the model), the higher the economy's growth path runs. The higher the depreciation rates of these stocks or the rate of growth of effective labour, the lower the analysed path of growth is positioned. Thirdly, investments in physical capital and human capital can be substituted, because a higher rate of investment in physical (or human) capital not only augments the stock of physical (human) capital per unit of effective labour, but also increases the other of the discussed stocks (cf. also relations (2.13a,b)), translating into improved productivity of labour and the economy's entering a higher-located path of growth.

2.4 Models of endogenous economic growth (from Ramsey to Lucas)

Models of endogenous economic growth that developed from the critical analysis of the Solow neoclassical model of growth give up the assumption about constant returns to scale of the aggregate production function and attempt to endogenize both technological progress and the economy-wide rate of saving/investment. The authors of the endogenous growth models believe that the Solow model of growth represented a considerable advancement compared with the Keynesian models of growth, because the former shows that a flexible capital-intensity ratio makes it possible to have a long-run economic growth without being permanently exposed to the knife-edge risk. Nevertheless, the exogeneity of both technological progress and the rate of saving/investment in the Solow model fall short of explaining why two real economies exhibit different rates of economic growth. Therefore, the endogenous models of growth abandon the two aforementioned assumptions and replace them with a hypothesis that the pace of accumulation of scientific and technical knowledge (in the Romer model [1986 and 1990]) or human capital (in the Lucas model [1988]) is determined by deliberate investments in these spheres. Moreover, the models assume that the rate of saving/investment implicitly takes a value maximizing the sum of discounted utilities of consumption of a typical entity or of the entire economy.

Moreover, the endogenous growth models *"take into account the occurrence of externalities related to knowledge, technological progress or capital. Investments (in both physical capital and human capital) lead then to productivity growth in excess of private benefits. If the externalities are strong enough to neutralise the impact of falling revenues, then a positive feedback between knowledge and investment can permanently shape the rate of growth."* (Wojtyna, [1996], p. 6). In other words, the endogenous growth models reject the neoclassical assumption about constant returns to scale of the aggregate production function.

Because most models of endogenous growth draw on the Ramsey model of growth [1928], either indirectly or directly, it seems practical to sketch the Ramsey model before the former are characterised. The Ramsey model assumes that (cf. also Barro, Sala-i-Martin, [1995], chapter 2; Roszkowska, [2004], pp. 31-42, etc.):

1. The process of production is described by a Cobb-Douglas production function with Harrodian technological progress given by the formula[2]:

[2] A standard Ramsey model assumes that a production function is a general, neoclassical function of production. Further analyses in this section will concentrate on the Cobb-Douglas production function, because the conclusions they provide are similar to those provided by discussions based on the general production function.

$Y=K^{\alpha}(AL)^{1-\alpha}$, whose variables and parameters have the same interpretation as in the Solow model.

2. These stocks of labour L and knowledge A expand at exogenous rates of growth of n and g ($n, g > 0$). Hence, $\dot{L}/L = n$ and $\dot{A}/A = g$, which implies that a stock of effective labour $\tilde{L} \equiv AL$ grows at the rate of $g + n$.
3. The analysed economy is perfectly competitive structure. Its enterprises work to maximize their profits. According to the marginal theory of division, this assumption means that the capital K and the labour L are rewarded in relation to their marginal products. Hence, the real wages w and the real interest rate r are given by the equations $w=\partial Y/\partial L$ and $r+\delta=\partial Y/\partial K$, where δ stands for the rate of capital depreciation.
4. Typical consumers in the Ramsey model (likewise in the subsequent models of endogenous growth) function to maximise the sum of discounted utilities over an infinite time horizon. The sum of the discounted utility of consumption is described by an improper integral (otherwise the integral of preference) of the form:

$$\int_{0}^{+\infty} \frac{c^{1-\sigma}-1}{1-\sigma} e^{-\rho t} dt \tag{2.14}$$

where c is consumption per employee (i.e. consumption of a typical consumer), $\sigma \in (0;1) \cup (1;+\infty)$ denotes inversion of the inter-temporal substitution of consumption, and $\rho > 0$ denotes a discount rate for a typical consumer.

5. In the Ramsey model, an increase in physical capital is given by the following differential equation:

$$\dot{K} = Y - C - \delta K \tag{2.15}$$

where $Y-C$ denotes, implicitly, the magnitude of investments carried out in the economy determined by the value of the accumulated savings.

The production function $Y=K^{\alpha}(AL)^{1-\alpha}$ and the differential equation (2.15) allow deriving the following relation:

$$\dot{k} = k^{\alpha} A^{1-\alpha} - c - (\delta + n)k \tag{2.16}$$

where $k \equiv K/L$ and $c \equiv C/L$/

Taking advantage of assumptions 4 and 5, the problem of maximising the integral of preference (2.14) under a restriction imposed by the equation of motion (2.16) can be written as the following Pontryagin's optimal control problem:

$$\left.\begin{array}{l} \max\limits_{c \geq 0} \int\limits_0^{+\infty} \frac{c^{1-\sigma}-1}{1-\sigma} e^{-\rho t} dt \\ \textit{with respect to:} \\ \dot{k} = k^\alpha A^{1-\alpha} - c - (\delta + n)k \\ \textit{and} \\ k(0) = k_0 > 0 \end{array}\right\} \tag{2.17}$$

where k_0 is the amount of physical capital per employee at time $t = 0$.

It can be shown (see, for instance, Tokarski [2005b], pp. 35-37) that the first order conditions necessary to solve the Pontryagin's optimal control problem (2.17) are in fact represented by the following simultaneous differential equations:

$$\left.\begin{array}{l} \dot{\tilde{k}} = \tilde{k}^\alpha - \tilde{c} - (\delta + g + n)\tilde{k} \\ \frac{\dot{\tilde{c}}}{\tilde{c}} = \frac{\alpha \tilde{k}^{\alpha-1} - (\rho + \delta + n + \sigma g)}{\sigma} \end{array}\right\} \tag{2.18}$$

where $\tilde{k} \equiv K/\tilde{L}$ and $\tilde{c} \equiv C/\tilde{L}$ are the stock of physical capital K and the stream of consumption C per unit of effective labour $\tilde{L}$, respectively. (2.18).

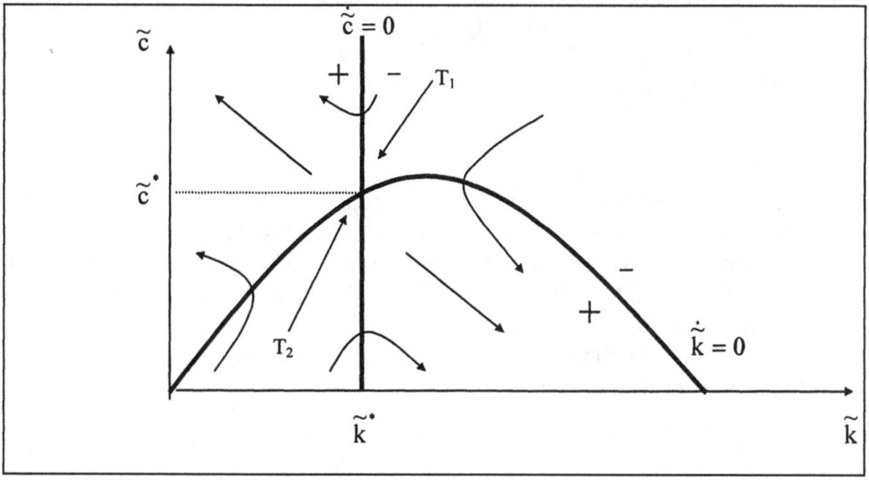

Figure 2.1.The phase diagram of the Ramsey model

According to the phase diagram, the Ramsey model has a saddle point. Therefore, if the Ramsey economy follows trajectory T_1 or T_2 at time $t=0$, then the economy tends to have a stable long-run equilibrium at the intersection of the demarcation curves $\dot{\tilde{c}}=0$ and $\dot{\tilde{k}}=0$. This implies that the conditions necessary for Pontryagin's maximum to exist are fulfilled (2.17). If, however, the Ramsey economy has its starting point outside the trajectories T_1 or T_2, then no stable position exists for the economy's long-run equilibrium (cf. also Roszkowska, [2004], pp. 41-32). Therefore, the likelihood of occurrence of stable Ramsey long-run equilibrium is as low as the likelihood that the Harrod-Domar economy is moving along the knife edge of the only growth path ensuring the economy long-run macroeconomic equilibrium.

The representative models of endogenous growth underpinned by the Ramsey model of growth seem to be those developed by Romer [1986; 1990] and Lucas [1988].

The Romer model is founded on the following assumptions:

1. The economy has a finite and invariable-in-time stock of human capital H. This capital comprises human capital allocated to production (H_Y) and to the sphere creating the scientific and technical knowledge (H_A). Hence:

   $$H=H_Y+H_A \tag{2.19}$$

2. The stock of technical and scientific knowledge varies in time as given by the following differential equation:

   $$\dot{A}=\kappa H_A A \tag{2.20}$$

   where $\kappa>0$ is a rate of effectiveness of human capital inputs used to create the scientific and technical knowledge.

3. The stream of product Y is described by an extended Cobb-Douglas production function of the form:

   $$Y=(H_Y)^{\beta}L^{\alpha}\int_0^A[x(i)]^{1-\alpha-\beta}di \tag{2.21}$$

   where $x(i)$ stands for the input of the i-th capital good $i\in(0;A)$; this means that the amount of capital goods (i) used in the production processes in the economy depends on the economy's stock of scientific and technical knowledge A. On the other hand, α, β and $(1-\alpha-\beta)\in(0;1)$ are the elasticities of Y with respect to L, H_Y and $x(i)$.

   With $\bar{x}=x(i)$ standing for an average capital good (for all i), the production function (2.21) can be written as follows:

   $$\begin{aligned}Y&=(H_Y)^{\beta}L^{\alpha}\int_0^A[x(i)]^{1-\alpha-\beta}di=(H_Y)^{\beta}L^{\alpha}[\bar{x}]^{1-\alpha-\beta}\int_0^A di=\\&=(H_Y)^{\beta}L^{\alpha}A[\bar{x}]^{1-\alpha-\beta}\end{aligned} \tag{2.22}$$

4. The total stock of physical capital in the economy is the sum of capital goods $x(i)$, which implies:

$$K = \int_0^A x(i)di = \bar{x}\int_0^A di = A\bar{x} \Rightarrow \bar{x} = K / A \tag{2.23}$$

An increase in the stock of physical capital $\dot{K}$ is the difference between production Y and consumption C (for the sake of simplicity, capital depreciation is omitted here). From this equation and from equations (2.21) - (2.23) it follows that:

$$\dot{K} = (H_Y A)^\beta (LA)^\alpha K^{1-\alpha-\beta} - C \tag{2.24}$$

5. The number of employees does not change in time and equals $L>0$.
6. The economy functions to maximise the sum of discounted utilities of consumption given by:

$$\int_0^{+\infty} \frac{C^{1-\sigma} - 1}{1-\sigma} e^{-\rho t} dt \tag{2.25}$$

where the economic interpretation of σ and ρ is the same as in the Ramsey model.

Based on the equations (2.19) - (2.24), the following equations of motion of the Romer model can be derived:

$$\left.\begin{aligned} \dot{A} &= \kappa H_A A \\ \dot{K} &= A^{\alpha+\beta}(H - H_A)^\beta L^\alpha K^{1-\alpha-\beta} - C \end{aligned}\right\} \tag{2.26}$$

The problem of finding the Romer model equilibrium consists in maximising the integral of preference (2.25) under a restriction imposed by the simultaneous differential equations (2.26). Using Pontryagin's maximum principle, we can show (cf. [Chiang, 1992, pp. 272-274]) that for the steady-state growth conditions (when $\dot{Y}/Y = \dot{K}/K = \dot{C}/C = \dot{A}/A = \kappa H_A$) the problem can be solved by the stock of human capital allocated to scientific and technical sphere as given by the formula:

$$H_A = \frac{\kappa(\alpha+\beta)H - \beta\rho}{\kappa(\beta\sigma+\alpha)}$$

and by the rates of economic growth of the form:

$$\frac{\dot{Y}}{Y} = \frac{\dot{K}}{K} = \frac{\dot{C}}{C} = \frac{\dot{A}}{A} = \frac{\kappa(\alpha+\beta)H - \beta\rho}{\beta\sigma+\alpha}$$

The above relations show that (under the additional condition $\kappa(\alpha+\beta)>\beta\rho$) both the magnitude of human capital allocated to the scientific and technical sphere and the rates of growth of the basic macroeconomic variables distinguished in the Romer model of economic growth are the higher, the higher the total stock of human capital in the economy (H) and the higher the exogenous rate of effectiveness of human capital inputs in the scientific and technological sphere (κ).

Moreover, the long-run rates of economic growth are also likely to be the higher, the more clearly households in the analysed economy give priority to future consumption over current consumption (i.e. the lower ρ and the higher σ).

Because the stock of human capital and its distribution between production and accumulation of knowledge essentially shapes the basic rates of economic growth in the Romer model, economies with a fairly low stock of this capital can be expected to show relatively small rates of growth. Another conclusion is that the integration and cooperation processes occurring between particular economies (involving both flows of human capital and externalities caused by the accumulation of scientific and technical knowledge) can be a factor equalizing the declining marginal products of the physical capital and increasing the rates of growth in the integrating economies (cf. also Romer, [1990], p. 71).

Another model of endogenous growth, now classical, is the Lucas model of 1988 that builds on the following assumptions (Lucas, [1988], pp. 7-8; [1990], [1993]; Tokarski, [1998]):

1. A typical and rational economic entity maximises the sum of discounted utilities of consumption per employee over an infinite period of time. This sum is given by the following integral of preference:
$$\int_0^{+\infty} \frac{c^{1-\sigma}-1}{1-\sigma} e^{-\rho t} dt \qquad (2.27)$$
where c is consumption per employee in the Lucas model and ρ and σ are interpreted as in the Ramsey model and Romer model.
2. The aggregate production function is given by the following formula:
$$Y = h^{\beta} K^{\alpha} (\Xi h l)^{1-\alpha} \qquad (2.28)$$
where α and $1\text{-}\alpha \in (0;1)$ are the elasticities of the stream of product Y with respect to inputs of the physical capital K and labour ΞhL in the production sphere, $\Xi \in (0;1)$ stands for the share of working time allocated to the production of goods and services, h is the stock of human capital in the economy, and $\beta \in [0;1)$ is the impact of so-called externalities generated by the capital accumulation processes (more on this subject in Tokarski ([1998], point 2).
3. An increase in the stock of physical capital is described by the differential equation:
$$\dot{K} = I - \delta K = Y - C - \delta K \qquad (2.29)$$
which has the same interpretation as the equation (2.15) in the Ramsey model.
4. An increase in the stock of human capital is described by the Uznawa-Rosen function as:
$$\dot{h} = \kappa(1 - \Xi)h \qquad (2.30)$$
where $\kappa > 0$ is the maximal, possible rate of growth of the stock of human capital.
5. The stock of labour grows at the rate of $n > 0$, which can be written as:

$\dot{L}/L = n$ (2.31)

6. The economy has constant rates of growth $g = \dot{y}/y = \dot{k}/k = \dot{c}/c$ and $g_h = \dot{h}/h$ (where $y \equiv Y/L$, $k \equiv K/L$ and $c \equiv C/L$) and a constant share of time allocated to production $\Xi(t) = \Xi^* \in (0;1)$ (for all $t \geq 0$).

Regarding the Lucas model of endogenous growth, the problem of finding the optimal path of economic growth consists in maximizing the integral of preference (2.27) under restrictions given by (2.28) – (2.31). The problem is resolved by means of Pontryagin's maximum principle. It can be demonstrated that under the assumptions 1-6, the optimal rates of growth, g and g_h, and the share of time allocated to production Ξ^* are given by the following formula:

$$g = \frac{1}{\sigma}\left(\frac{1-\alpha-\beta}{1-\alpha}\kappa - \rho\right),$$

$$g_h = \frac{1}{\sigma}\left(\kappa - \frac{1-\alpha}{1-\alpha+\beta}\rho\right)$$

and:

$$\Xi^* = 1 - \frac{1}{\sigma}\left[1 - \frac{(1-\alpha)\rho}{(1-\alpha+\beta)\kappa}\right]$$

which means, *inter alia*, that for $\frac{1-\alpha+\beta}{1-\alpha} \geq \frac{\rho}{\kappa} \geq \frac{1-\alpha+\beta}{1-\alpha}(1-\sigma)$ the following equations are true:

$$g = g\left(\overset{+}{\frac{1}{\sigma}}, \overset{-}{\rho}, \ldots\right) \tag{2.32}$$

$$g_h = g_h\left(\overset{+}{\frac{1}{\sigma}}, \overset{-}{\rho}, \ldots\right) \tag{2.33}$$

From the equations (2.32 - 2.33) it follows that the rates of growth for product streams, consumption, the stock of physical capital (all per employee), as well as individual qualifications of employees mainly depend on factors describing the structure of consumption preferred in time (i.e. on $1/\sigma$ and ρ). The more clearly the typical consumers in the Lucas model favour current consumption over future consumption (i.e. the smaller σ or the larger ρ), the lower the rate of growth the studied economy can have. Moreover, unlike the Solow or M-R-W neoclassical models, in the Lucas model of endogenous growth the long-run rate of economic growth can be increased permanently owing to externalities generated by the accumulation of human capital. However, the preferred structure of consumption must change in time for this increase to take place.

Another finding arising from the Lucas model is that the poor economies (with low stocks of physical and human capital per employee) have slim chances of catching up with the better-off nations without changing in time their preferred structures of consumption, which is not easy to accomplish. Furthermore, even if the possible flows of physical capital between particular economies (including its flows from the rich countries to the poor countries) are taken into account, this does not basically change the above statement. Therefore, the poor countries (having low stocks of k and h) do not have to have significantly higher marginal products of physical capital than the rich countries (where the stocks are substantial). From this it follows that the rich and poor countries have similar rates of return on physical capital, and that an economically justifiable reason for physical capital to flow from the rich countries to the poor ones does not exist (cf. also Lucas, [1990], pp. 92-94 and 96).

2.5 Summary and conclusions

The above discussion can be summed up as follows:

1. The modern theories of economic growth started to develop with the short-run analyses of factors determining the magnitude of the product in the Keynesian multiplier model and Hicks IS-LM model. These models accentuate only the short-term impacts of investment outlays on the value of aggregated demand and the level of product generated (basically) under underutilised production capacity. Unlike the Keynesian models, the post-Keynesian models of growth address also the long-run effects of investment outlays on the supply side of the economy.
2. The post-Keynesian models of economic growth attempted to identify the conditions under which the supply-side and demand-side effects of investment outlays would balance out when the economy's capacity were fully utilised. The main conclusion that the models provide is that the economy is almost constantly pushed to deviate from a path of growth guaranteeing macroeconomic equilibrium (the so-called knife-edge problem). Besides, if the state does not decide to intervene actively, a single deviation from this path triggers a building-up disequilibrium in the product market and, implicitly, in the labour market.
3. Although fluctuations are part of economies' functioning, the economic cycles seem to be self-limiting, which implies that the economies do not fall off the knife's edge of the Keynesian growth models. Based on this fact, Solow concluded that the analysis of the neoclassical production functions could provide a better explanation of the processes behind the long-run economic growth than the examination of the supply and demand adjustments. He also demonstrated that economies naturally gravitate to the long-run path of steady-state growth. On the long-run path of growth, the rate of economic growth is determined by the rate of exogenous technological progress. The Solow model additionally indicates that a change in the rate of saving/investment only

moves the economy from its original path of steady-state growth to a path running either above or below the original one, but does not permanently affect the long-run rate of economic growth.

4. The Mankiw-Romer-Weil model draws on the Solow model in the sense that the economy it models has constant returns to scale with respect to the inputs of all production factors. However, it must be treated as an extension of the Solow model, as it adds the stock of human capital to the arguments of the macroeconomic production function. This stock affects the production processes in the same way as the stocks of physical capital and effective labour do, and human capital is accumulated because of organizations' intended investments in this sphere (although the Solow model treats the rates of investment in human and physical capital and the rate of saving/investment as exogenous variables). The main conclusions arising from the Mankiw-Romer-Weil model are the following. Firstly, the economy naturally gravitates to a rate of growth determined by the rate of exogenous technological progress. Secondly, the paths of growth along which the Mankiw-Romer-Weil economy moves can change following changes in the rates of investment in the stocks of physical and human capital. Thirdly, the higher the rates of investment in the capital stocks distinguished in the Mankiw-Romer-Weil model, the higher the long-run paths of economic growth are located.
5. Another theory trying to explain the determinants of economic growth is theory of endogenous growth. The leading representatives of this school start with critical analysis of the Solow model of growth and maintain that a model of economic growth can be constructed, where the rate of technological progress and the rates of saving/investment are endogenised, and the assumption of constant returns to scale of the production function is abandoned. The authors of endogenous growth theory assume that technological progress (understood as the accumulation of human capital or scientific and technical knowledge) is powered by organizations' intended investments in these spheres. Moreover, in line with the major points formulated by new classical economy and the school of real economic cycle, the rate of saving/investment should have a value maximizing the sum of discounted utilities of a typical entity in the economy. The assumption about constant returns to scale of the macroeconomic production function is rejected, because accumulation of technical and scientific knowledge and human capital generates externalities. Externalities appear because the accumulated knowledge and human capital can be used not only by organizations that actually pay the accumulation costs, but also by entities in their micro and/or macroeconomic environment.
6. The major conclusion that endogenous growth theories allow drawing is that owing to the existence of externalities (related to technological progress) the long-run rate of economic growth can be increased effectively and

permanently. For this to happen, however, a relatively high priority must be given to future consumption, on one hand, and investments promising high rates of return (in technical and scientific knowledge and human capital) need to be emphasised, on the other.

References:

Barro R.J., Sala-i-Martin X., 1995, *Economic Growth,* McGraw-Hill Inc., New York.

Chiang A.C., 1992, *Elements of Dynamic Optimization,* McGraw-Hill International Editions, New York.

Chiang A.C., 1994, *Podstawy ekonomii matematycznej, (Basic Mathematical Economics)* PWE, Warszawa.

Domar E.D., 1962, *Szkice z teorii wzrostu gospodarczego,* (Essays in Economic Growth Theory) PWN, Warszawa.

Górski J., Sierpiński W., 1972, *Historia powszechnej myśli ekonomicznej 1870-1950 (History of Economic Thought 1870-1950),* PWN, Warszawa.

Harcourt G.C., 1975, *Spory wokół teorii kapitału,* (Dillemas around the Theory of Capital), PWE, Warszawa.

Harrod R.F., 1939, *An Essays in Dynamic Theory,* „Economic Journal", vol. 49.

Harrod R.F., 1942, *Toward a Dynamic Economics.* Macmillan, London.

Kaldor N., 1971, *Eseje z teorii stabilizacji i wzrostu gospodarczego. (Essays in Theory of Stabilization and Economic Growth*), PWN, Warszawa.

Kalecki M., 1963, *Zarys teorii wzrostu gospodarki socjalistycznej. in*: M. Kalecki, 1984, *Dzieła,* t. 4, *Socjalizm. Wzrost gospodarczy i efektywność inwestycji,* PWE, Warszawa.

Kalecki M., 1970, *Teorie wzrostu w różnych systemach społecznych.* (Theory of Growth for Different Social Systems) *in*: M. Kalecki, 1984, *Dzieła,* t. 4, *Socjalizm. Wzrost gospodarczy i efektywność inwestycji,* PWE. Warszawa.

Lucas R.E., 1988, *On the Mechanics of Economic Development,* „Journal of Monetary Economics", vol. 22.

Lucas R.E., 1990, *Why Doesn't Capi ted Flow from Rich to Poor Countries?,* „American Economic Review", vol. 80.

Lucas R.E., 1993, *Making a Miracle,* „Econometrica", vol. 72.

Mankiw N.G., Romer D., Weil D.N., 1992, *A Contribution to the Empirics of Economic Growth.* „Quarterly Journal of Economics", vol. 107.

Phelps E.S., 1961, *The Golden Rule of Accumulation: A Fable for Growthmen.* „American Economic Review", vol. 54.

Ramsey F., 1928, *A Mathematical Theory of Saving,* „The Economic Journal", vol. 38.

Romer D., 1996, *Advanced Macroeconomics,* McGraw-Hill Inc., New York.

Romer P.M., 1986, *Increasing Returns and Long-Run Growth.* „Journal of Political Economy", vol. 94.
Romer P.M., 1990, *Endogenous Technological Change.* „Journal of Political Economy", vol.98.
Roszkowska S., 2004, *Inwestycje a wzrost gospodarczy w krajach OECD, (Investments and Economic Growth in OECD Countries*), praca magisterska napisana w Katedrze Ekonomii Uniwersytetu Łódzkiego pod kierunkiem T. Tokarskiego, Łódź.
Solow R.M., 1956, *A Contribution to the Theory of Economic Growth,* „Quarterly Journal of Economics", vol. 70.
Solow R.M., 1957, *Technical Change and the Aggregate Production Function,* „Review of Economics and Statistics", vol. 39.
Solow R.M., 1979, *Another Possible Source of Wage Stickiness,* „Journal of Macroeconomics", vol.1
Solow R.M., 1988, *Growth Theory and After,* „American Economic Review", vol. 79.
Summers L.H., 1988, *Relative Wages, Efficiency Wages, and Keynesian Unemployment,* „American Economic Review", vol. 78.
Swan T.W., 1956, *Economic Growth and Capital Accumulation.* „Economic Record", vol. 32.
Tokarski T., 1998, *Postąp techniczny a wzrost gospodarczy w modelach Solowa i Lucasa* (*Technical Progress and Economic Growth in Solow and Lucas Models*) , „Ekonomista", nr 2-3.
Tokarski T., 2003a, *On the Long-Run Monetary Policy Rules in the Domar-Solow Type of Economy. in*: W. Welfe, A. Welfe, 2003, *Macromodels 2002. Proceedings of the Twenty Ninth International Conference Macromodels,* University of Lodz, Łódź.
Tokarski T., 2003b, *Wzrost gospodarczy a rynek pracy w neoklasycznych modelach wzrostu,* (*Economic Growth and Labour Market in Neoclassical Models of Growth*), „ Studia Ekonomiczne" INE-PAN, vol. 38, nr 3.
Tokarski T., 2003c, *Specyfikacja funkcji produkcji a równowaga długookresowego wzrostu gospodarczego* (*Specification of Production Function and Long-term Economic Growth Equilibrium*), „Ekonomista", nr 3.
Tokarski T., 2005a, *Wybrane modele podażowych czynników wzrostu gospodarczego* (*Models of Supply Determinants of Economic Growth*), Wydawnictwo Uniwersytetu Jagiellońskiego, Kraków.
Tokarski T., 2005b, *Stopy procentowe a wzrost gospodarczy.* (*Interest Rates and Economic Growth*), „ Studia Ekonomiczne", INE-PAN, vol. 44-45, nr 1-2.
Welfe W. (ed.), 2004, *Długookresowy, makroekonometryczny model W8D-2002 gospodarki polskiej* (*Long-term Macroeconometric Model W8D-2002 of the Polish Economy*), „Acta UŁ, Folia Oeconomica", vol. 172, Wydawnictwo Uniwersytetu Łódzkiego, Łódź.

Wojtyna A., 1996, *Rola państwa we wzroście gospodarczym,* (*Role of the Goverment in Economic Growth*) paper presented at conference „Współczesne teorie wzrostu gospodarczego", PTE, Warszawa, październik 1996.

Chapter 3

Macroeconometric models of the knowledge-based economy

3.1 Introduction

A macroeconometric modelling of national economies has a long tradition. An outline of the world history of macroeconometric modelling (to the year 1990) has been provided in the monograph by Bodkin, Klein, Marwah [1991], whereas W. Welfe [1992] characterized Poland and other East and Central European countries. Developments that followed in the next years were reported on in monographs authored by Whitney [1994], Klein, A. Welfe, W. Welfe [1999], W. Welfe, A. Welfe [2004]. The most recent research activities aimed at applying cointegration analysis to the process of constructing small, monetary policy-oriented models can be found in the monographs by Bärdsen et al. [2005] for the Norwegian economy model and by Garratt et al. [2006] for the UK model.

The long-term macroeconometric models, extended to include processes in which knowledge capital is generated and used, seem to be the most relevant tools of long-term economic analysis (Welfe, [2000]). Their structure should follow the framework of the mainstream models outlined by Klein et al. [1999].

These models specify the final demand equations along the neo-Keynesian lines, but the potential output and demand for the factors of production, as well as impacts of technological progress they generate refer to the neoclassical theory of production (Solow, 1957). This approach draws on the early theories of growth developed by Harrod and Domar and on the concept of models of production possibility frontier that have recently been developed by Jorgenson [2000]. The stylized empirical model of growth by W. Welfe [2005] follows a similar approach.

From the above concepts it follows that the models of a knowledge-based economy should incorporate two basic blocks of equations – one explaining domestic final demand and foreign trade and another accounting for the supply of the production factors and impacts of technological progress and potential output. The models close with blocks of equations explaining prices and wages, as well as financial flows[1] (Welfe, [2008]).

Such decomposition of the national economy has an obvious economic justification. It seems more appropriate than one based on the microeconomic principles and distinguishing equations explaining the behaviour of households, producers, financial institutions, etc. However, the use of cointegration techniques of parameter estimation seems to suggest still another decomposition[2].

[1] Detailed discussions of the structure of macroeconometric models and specifications of particular equations for Poland can be found in W. Welfe, A. Welfe [2004] and W. Welfe ed. [2001, 2004].

[2] Please note that when the macromodels were first used the equations were rearranged to arrive at an almost recursive model structure.

Most macroeconometric, operational models are medium-sized (with 100-300 equations) and they are estimated using size-constrained samples (spanning 15-30 years or 40-60 quarters). For this reason, the single equation estimation techniques were mainly used in the past, and in the case of the non-stationary variables this could produce the spurious regressions. To prevent this risk, methods assuming cointegrated relations and vector error correction were developed. However, this approach involves the construction of small submodels (4-9 equations) explaining the cointegrated variables. Then, a special type of decomposition must be applied to the large macroeconometric models. It becomes necessary to distinguish blocks of equations whose variables are strongly correlated within a group, but weakly related to variables in other groups of equations. The problem of combining the submodels into a constituent system has not been resolved yet (see Bördsen et al. [2005]). Garratt et al. [2006] advocate a top-down approach, where the satellite models can surround the small core models.

3.2 Modelling final demand

Total final demand comprises domestic final demand H_t and net foreign demand EM_t. It determines demand for the domestic output X_t

$$X_t = H_t + EM_t \tag{3.1}$$

The domestic final demand can be decomposed into household (consumer) demand C_t, demand of public institutions G_t, demand for investments in fixed capital J_t, and demand for inventory increase ΔR_t:

$$H_t = C_t + G_t + J_t + \Delta R_t \tag{3.2}$$

Proposals have been formulated in the recent years to extend the notion of demand for investment goods so that investments in knowledge capital, i.e. R&D and educational expenditures, can be included, in addition to investments in fixed capital[3]. If such an extension were accepted, a pertinent modification of the SNA would be required, reducing the demand for consumer goods of public institutions and increasing total investments[4].

The modelling of household demand for consumer goods (individual consumption) has a long-standing tradition. When the knowledge-based economies are being modelled, it is necessary to emphasise specifications that are relevant in the long run. Hence, the theoretical point of departure should be F. Modigliani's life-cycle hypothesis. Under this hypothesis, consumers shape their long-run demand in accordance with the real disposable income they expect

[3] Zienkowski [2003] suggests to introduce a new macro category called investment in development.

[4] Jorgenson adds the value of services provided by dwellings and consumer durables to the private sector's GDP (Jorgenon et al. [2002]).

for the lifetime and accumulated savings. In many countries, however, for instance Poland, this applies to the behaviour of a rather small fraction of the population. Most households living in unstable environments are constrained by either real disposable income Y_t or credit. Their behaviour is additionally determined by the incomes they derived in the past, SV_t, and current interest rates. Hence, the long-run demand function for consumers reads:

$$C_t = AY_t^{\alpha} SV_t^{(1-\alpha)} RD_t^{\beta} \varepsilon^{\varepsilon_t}, \qquad (3.3)$$

where

RD_t - interest rate on deposits.

In the course of estimation based on an ECM model, the relevant lags and additional variables need to be incorporated in order to allow for the short-run adjustments (W. Welfe, A. Welfe [2004]).

The main determinant of public institutions' demand for consumer goods is current, real expenditures of the government – BC_t. A more adequate specification would take as the explaining variable the current budget expenditures from which the public debt service and government subsidies to the private sector were deducted.

The demand function reads as follows:

$$G_t = A(BC_t)^{\alpha} e^{\varepsilon_t} \qquad (3.4)$$

The short-term adjustments are typically represented by lags related to the infrastructure expenditures.

Specification of private enterprises' demand for investment goods starts with an accelerator. To put it simply, producer capacities expand in the long run following expected increase in the demand for products the producers can provide. This increase creates, allowing for restitution demand (and a likely improvement of the level of utilization of the available equipment), the potential demand for investment goods. The perception of investment risks and changes in the profitability of investment projects also has an effect on the effective demand.

A typical representation of the expected future output is output generated in the past. As for machinery and equipment, the Koyck transformation leading to a reduced form of the investment demand function can be used, where the explanatory variables are being confined to lagged investment (J_{t-1}) and current output X_t (see W. Welfe, A. Welfe [2004]). The investment risk RJ_t can be approximated using various indicators, such as government deficit, government debt service, or a rate of inflation. Profitability of investment projects is usually expressed by the ratio of producer prices (PX_t) and user costs KI_t, the latter predominantly depending on the long-term interest rate RL_t. Hence, a typical specification of the investment demand function is as follows:

$$J_t = AJ_{t-1}^{\alpha_1} X_t^{\alpha_2} RJ_t^{\alpha_3} (PX_t / KI_t)^{\alpha_4} e^{\varepsilon_t} \qquad (3.5)$$

where
$KI_t = PJ_t(RL_t + \delta)$, PJ_t – an investment deflator,
RL_t – real long-term interest rate,
δ – the rate of depreciation.

The above investment function can be enlarged even further, for instance to reflect changes in the tax rates.

An increase in inventories is most frequently related to output increase, taking into account the initial level of inventories:

$$\Delta R_t = \alpha_o + \alpha_1 \Delta X_t - \alpha_2 R_{t-1} + \varepsilon_t \qquad (3.6)$$

If accurate data are available, the specification is extended by adding variables that approximate the differences between the estimated demand D and the supply S that usually represent the level of capacity utilization (i.e. the output gap).

Equations explaining foreign trade can be constructed in a traditional manner. Imports and exports depend on the levels of economic activity and relative prices.

The demand for imported commodities M_t is shaped by GDP. Total imports are frequently decomposed with reference to the use of imported commodities. Imports of consumer goods, mainly of the competitive ones, are determined by total domestic consumption, and imports of investments goods follow domestic investments. The demand for intermediate goods rests on domestic output.

The demand for imported substitutes is also determined by the relative prices of imported commodities, denominated in either USD or EURO, with respect to domestic prices.

Hence, at the macro level, we have

$$M_t = BX_t^{\alpha}(PM_t * WZLD_t / PX_t)^{\alpha_2} e^{\varepsilon_t}, \qquad (3.7)$$

where $WZLD_t$ is the Polish zloty/USD exchange rate.

The "income" elasticity α_1 is frequently calibrated at the level of α_1=1. The underlying assumption that the import/GDP ratio is constant in the long run does not seem valid, as the spectrum of commodities entering the world trade seems to expand in time.

The specification of exports equations is largely alike. The demand for exported commodities E_t is equal to total demand for imports in the countries receiving the exported goods. Hence, the buying countries' imports or even total world imports, MW_t, are assumed to determine exports. The substitution effects are dealt with by including the relative export prices into equations explaining exports. The relative export prices are calculated as ratios of the domestic prices PX_t (converted into USD or Euro) to the world prices PED_t.

$$E_t = B(MW)_t^{\alpha_1}(PX_t / WZLD_t / PED_t)^{\alpha_2} e^{\varepsilon_t} \qquad (3.8)$$

Sometimes, the impact of the relative prices is narrowed to domestic sectors producing exportables.

3.3 Modelling the process of production

Domestic products are generated via the process of production that transforms the factors of production – fixed capital and labour – by means of a specific technology and organization of labour. Analysis of the production process requires the quantification of its specific relationships. An important instrument used in such analysis is the production function.

Among its various forms, a double-log function and a CES function with constant elasticity of substitution are frequently used. Following many authors, we recommend the double-log production function, i.e. a Cobb-Douglas function with constant returns to scale in its extended version:

$$X_t^P = BA_t K_t^{\alpha} N_t^{(1-\alpha)} e^{\varepsilon_t} \qquad (3.9)$$

where

X_t^P – potential output (GDP at the macroscale) in constant prices

A_t – total factor productivity (TFP),

K_t – fixed capital, constant prices,

N_t – employment.

Estimating the function's parameters for a market economy is not a straightforward exercise, because observations represent effective output, i.e. the realization of the final demand X_t, and not potential output, and most frequently $X_t < X_t^P$. Therefore, changes in the rate of utilization of production potential must be addressed. In most cases, they originate from changes in final demand, i.e. from the business cycle fluctuations. The rate of capacity utilization is defined as

$$WX_t = X_t / X_t^P \qquad (3.10)$$

The rate can be decomposed further into the rates of utilization of the fixed capital WK_t and total time worked WN_t. Owing to these partial rates of utilization, we can redefine the production function (3.9) into:

$$X_t = BA_t (K_t WK_t)^{\alpha} (N_t WN_t)^{(1-\alpha)} e^{\varepsilon_t} \qquad (3.11)$$

or

$$X_t = BWX_t A_t K_t^{\alpha} N_t^{(1-\alpha)} e^{\varepsilon_t} \qquad (3.11')$$

where $WX_t = WK_t^{\alpha} WN_t^{(1-\alpha)}$,

Taking logarithms on both sides of equation (3.11), we obtain:

$$\ln X_t = \ln B + \ln A_t + \alpha \ln(K_t WK_t) + (1-\alpha)\ln(N_t WN_t) + \varepsilon_t \qquad (3.12)$$

To estimate the parameters of the above function, we need information on the rate of factor utilization and total factor productivity. Seeking a relevant approximation of the utilization rates of particular factors is not a trivial task, because the rates differ in terms of their accuracy and statistical coverage. Some methods take advantage of industrial surveys that ask direct questions about the level of utilization of machinery and equipment, as well as time worked (Grzęda-Latocha [2005]). Additionally, central banks use techniques involving analysis of the residuals (deviations from a GDP trend). There are also procedures using the aforementioned decomposition of the rate of utilization of fixed capital and employment. Because the first category fails to provide empirical estimates, the characteristics of utilization of time worked by employees are usually used. Several countries use total time worked by employees as the explanatory variable, which can be decomposed into the number of shifts worked and the length of the working time. These characteristics can serve as the approximate measures of time worked by machinery and equipment (see W. Welfe [1992]).

The level of TFP, and especially its growth rates, is not directly observable. As we have already mentioned in the previous chapters, the most popular approach to estimating TFP values is based on Solow residual. The rate of TFP growth will be calculated by subtracting the rate of growth of potential GDP, X_t^*, obtained from the production function while ignoring the impact of TFP, i.e. assuming that $A_t = 0$, from the rate of growth of effective, GDP X_t. Then we have:

$$\Delta \ln A_t = \Delta \ln X_t - \Delta \ln WX_t - \alpha \Delta \ln K_t - (1-\alpha)\Delta \ln N_t \qquad (3.13)$$

The value of elasticity, α, is either calibrated or estimated. In the latter case, the estimate of parameter α and parameters entering the function explaining TFP, i.e. A_t, must be calculated and estimated respectively at the same time.

Empirical research frequently neglects the term representing the capacity utilization (WX_t), which clearly leads to biased estimates of TFP dynamics (Welfe, [2002]).

Endogenisation of technological progress implies that the equation explaining TFP will be specified. Early investigations commonly assumed exogeneity of TFP, i.e. its being an exponential function of time. Endogenisation of technological progress led to the conclusion that the dynamics of TFP can be explained by means of an increase in the capital of knowledge. Knowledge capital is either freely disseminated among the economic agents, or represented by the results of cumulative R&D expenditures, both domestic, BRK_t^K, and transferred from abroad, BRK_t^M, and by human capital per employee h_t. In general, we have:

$$\ln A_t = \alpha_0 + \beta_1 \ln BRK_t^K + \beta_2 \lambda \ln BRK_t^M + \gamma \ln h_t + \varepsilon_t \qquad (3.14)$$

Specification of the above function is presented in detail the next chapters.

3.4 Price and wage equations. Modelling the financial flows

Producer prices (GDP deflator at the macroscale PX_t) are central to the modelling of a price system. They can be split up into the unit costs KIP_t and the mark-up AF_t. The unit costs can be further decomposed into unit labour costs that depend on average wages and productivity of labour, costs of intermediate inputs (both domestic and imported, shaped by the world prices), depreciation and other unit costs.

The main reason behind the frequent use of the reduced-form equations of producer prices is the scarcity of the data on the use of intermediate commodities. Especially in the long run, the labour costs KW_t and the impact of the imported inflation PM_t are the only explanatory variables available. Assuming homogeneity, we have:

$$PX_t = B(AF_t)^{\alpha}(KW_t)^{\beta}(PM_t)^{(1-\alpha-\beta)}e^{\varepsilon_t} \qquad (3.15)$$

where

$AF_t = f(WX_t)$, a mark-up depending on the rate of capacity utilization,

$KW_t = WBP_t(1+\upsilon_t): N_t / X_t$, where WBP_t is average wage and υ_t is a mark-up on wages,

$PM_t = PW_t * WZDL_t$, where PW_t is the world prices in USD and $WZLD_t$ is the zloty/USD exchange rate.

In the short-run, producer prices fluctuate following changes in the inflationary expectations that are in turn induced by changes in producer prices that occurred in the past periods.

As a rule, the deflators of particular components of final demand and fixed capital are generated from the bridge equations. Their variations are caused by changes in the producer prices and prices of imports, and/or by changes in the indirect tax rates, depending on the situation[5].

Equations explaining average wages were the subject of many investigations in the past. The investigations' point of departure was analysis of the relationship between the rate of growth of average wages and the rate of unemployment (Phillips curve). Another important factor was the rate of inflation. Recently, though, the low rate of inflation has made researchers assume that the wage rates grow proportionally to the rate of inflation. Consequently, the real average wages WB_t are explained instead of nominal wages WBP_t. It is believed that real average wages are determined in the long run by the productivity of labour (X_t / N_t), but the assumption (calibration) that the value of their elasticity

[5] Equations explaining the prices of exports are structured in a special way. They are formed as a combination of the world prices (PW_t) and domestic prices (PX_t). As a result, the impacts of price policies can be translated into the changes of export prices.

is 1 is quite unrealistic. Of course, average wages also depend on the rate of unemployment U_t. We then have

$$\Delta \ln WB_t = \alpha_o + \alpha_1 \Delta \ln(X_t / N_t) + \alpha_2 / U_t + \varepsilon_t \qquad (3.16)$$

Both labour efficiency and the impact of unemployment are adjusted in the short-run, also because of changing inflationary expectations.

Prices functioning in the financial markets (interest rates, etc.) were treated in the past as exogenous. Recently, however, a common tendency to endogenize these prices has been observed, with the central bank interest rates being viewed as a major instrument of monetary policy. This instrument, used by the central banks and/or their advisory boards, is changing as a result of widespread discussions. The institutions strive for stabilizing inflation or, additionally, for supporting economic growth, but their decisions are somewhat arbitrary.

Nevertheless, the knowledge about their behaviour in the past corroborates the hypothesis that the current interest rate is affected by the earlier variations in the rate of inflation. Hence, this relationship can be approximated as follows:

$$R_t = f[(PX_t - PX_{t-1}) / PX_{t-1}] \qquad (3.17)$$

The interest rates that the commercial banks charge on loans and pay on deposits are derived from the bridge equations.

In an open economy, the exchange rate is an important determinant of changes in the balance of payments, and the PPP hypothesis is central to its specification. In the short-run, however, the likely deviations from the basic exchange rate are explained using the flows of financial capital. According to the UIP hypothesis, the level of the flows usually depends on the differences between interest rates operated in the compared countries. For example, the exchange rate of USD vis-à-vis Polish Zloty may have the following form:

$$WZLD_t = A(PX / PW_t)^{\alpha} (R_t / R_t^W)^{\beta} e^{\varsigma_t} \qquad (3.18)$$

where R_t^W is the foreign interest rate.

3.5 Modelling based on non-stationary time series [6]

Let us assume that a data-generating process (DGP) contains a stochastic and deterministic component:

$$\mathbf{v}_{(m)t} = \mathbf{y}_{(m)t} + \mathbf{d}_{(z)t}\mathbf{H}, \qquad (3.19)$$

where:

$\mathbf{y}_{(m)t} = [y_{1t} \cdots y_{Mt}]$ - the 1xM vector M of I(1) variables,

$\mathbf{d}_{(z)t} = [d_{1t} \ldots d_{Zt}]$ - the 1xZ vector Z of deterministic variables,

[6] For details see A. Welfe [2009].

$$\mathbf{H} = \begin{bmatrix} \mathbf{h}_{1(m)} \\ \vdots \\ \mathbf{h}_{Z(m)} \end{bmatrix}$$ - the ZxM matrix of parameters of deterministic variables,

$m = 1,\ldots,M$; $z = 1,\ldots,Z$.

In most cases, the list of the deterministic variables includes only the constant and the trend, while the structural changes are reflected by changes in the relevant parameters. Then $\mathbf{d}_{(z)t} = [1 \vdots t \vdots \mathbf{u}_{1(w)} \vdots t\mathbf{u}_{2(w)t}]$, and the vectors $\mathbf{u}_{1(w)t}$ and $\mathbf{u}_{2(w)t}$ of dimensions 1xW are composed of dummy variables. Let us assume that a structural change occurred in one period only, κ; then we have:

$$\mathbf{v}_{(m)t} = \mathbf{y}_{(m)t} + \mathbf{h}_{1(m)} + t\mathbf{h}_{2(m)} + u_t\mathbf{h}_{3(m)} + tu_t\mathbf{h}_{4(m)}, \tag{3.20}$$

where $u_t = \begin{cases} 1 & \text{dla} \quad t \geq \kappa \\ 0 & \text{dla} \quad t < \kappa \end{cases}$.

Multiplying both sides of the above equation by $\mathbf{\Pi}(L)$, where $\mathbf{\Pi}(L) = \mathbf{I} - L\mathbf{\Pi}_1 - L^2\mathbf{\Pi}_2 + \ldots - L^S\mathbf{\Pi}_S$, and $\mathbf{\Pi}_S$ (s=1,..,S) denote MxM matrices of parameters, we arrive at:

$$\mathbf{v}_{(m)t}\mathbf{\Pi}(L) = \mathbf{y}_{(m)t}\mathbf{\Pi}(L) + \mathbf{h}_{1(m)}\mathbf{\Pi}(L) + t\mathbf{h}_{2(m)}\mathbf{\Pi}(L) + u_t\mathbf{h}_{3(m)}\mathbf{\Pi}(L) + tu_t\mathbf{h}_{4(m)}\mathbf{\Pi}(L) \tag{3.21}$$

Let us assume that the order of the cointegrated variables y_{mt} is R, $R \in (0, M)$, then for the VAR model $\mathbf{v}_{(m)t}\mathbf{\Pi}(L) = \boldsymbol{\xi}_{(m)t}$, where $\boldsymbol{\xi}_{(m)t} = [\xi_{1t} \ldots \xi_{Mt}]$ is 1xM vector of white-noise errors, a VEqCM exists:

$$\Delta\mathbf{v}_{(m)t} = \mathbf{v}_{(m)t-1}\mathbf{\Pi} + \sum_{s=1}^{S-1} \Delta\mathbf{v}_{(m)t-s}\mathbf{\Gamma}_s + \boldsymbol{\xi}_{(m)t}, \tag{3.22}$$

where:

$\mathbf{\Pi} = \sum_{s=1}^{S} \mathbf{\Pi}_s - \mathbf{I}$,

$\mathbf{\Gamma}_i = -\sum_{s=i+1}^{S} \mathbf{\Pi}_s = -(\mathbf{\Pi}_{i+1} + \mathbf{\Pi}_{i+2} + \ldots + \mathbf{\Pi}_S)$,

$\mathbf{\Pi} = \mathbf{B}\mathbf{A}^T$,

B - the MxR matrix of cointegrating vectors,

A - the MxR matrix of weights.

Analogously, the representation of VAR for $\mathbf{y}_{(m)t}$ can be written as:

$$\mathbf{y}_{(m)t}\mathbf{\Pi}(L) = \boldsymbol{\xi}_{(m)t} \,. \tag{3.23}$$

The first two deterministic components on the right-hand side of the equation (3.21) are respectively given by:

$$\mathbf{h}_{1(m)}\mathbf{\Pi}(\mathbf{L}) = \mathbf{h}_{1(m)}(\mathbf{I} - \mathbf{L}\mathbf{\Pi}_1 - \mathbf{L}^2\mathbf{\Pi}_2 - \ldots - \mathbf{L}^S\mathbf{\Pi}_S) = -\mathbf{h}_{1(m)}\mathbf{B}\mathbf{A}^T \,, \tag{3.24}$$

$$t\mathbf{h}_{2(m)}\mathbf{\Pi}(\mathbf{L}) = \mathbf{h}_{2(m)}[t\mathbf{I} - (t-1)\mathbf{\Pi}_1 - (t-2)\mathbf{\Pi}_2 - (t-3)\mathbf{\Pi}_3 - \ldots - (t-S)\mathbf{\Pi}_S] =$$

$$= \mathbf{h}_{2(m)}[(t-1)\mathbf{I} - (t-1)\mathbf{\Pi}_1 - (t-1)\mathbf{\Pi}_2 - (t-1)\mathbf{\Pi}_3 - \ldots - (t-1)\mathbf{\Pi}_S + \mathbf{I} + \mathbf{\Pi}_2 + 2\mathbf{\Pi}_3 + \ldots +$$

$$+ (S-1)\mathbf{\Pi}_S] = (t-1)\mathbf{h}_{2(m)}(\mathbf{I} - \mathbf{\Pi}_1 - \mathbf{\Pi}_2 - \mathbf{\Pi}_3 - \ldots - \mathbf{\Pi}_S) + \mathbf{h}_{2(m)}(\mathbf{I} + \mathbf{\Pi}_2 + 2\mathbf{\Pi}_3 + \ldots +$$

$$+ (S-1)\mathbf{\Pi}_S) = -(t-1)\mathbf{h}_{2(m)}\mathbf{B}\mathbf{A}^T + \mathbf{h}_{2(m)}\mathbf{\Psi} \tag{3.25}$$

where $\mathbf{\Psi} = \mathbf{I} - \sum_{s=1}^{S-1}\mathbf{\Gamma}_s = \mathbf{I} + \sum_{s=1}^{S-1} s\mathbf{\Pi}_{s+1}$.

The third component can be transformed as follows:

$$u_t\mathbf{h}_{3(m)}\mathbf{\Pi}(L) = u_t\mathbf{h}_{3(m)} - u_{t-1}\mathbf{h}_{3(m)}\mathbf{\Pi}_1 - u_{t-2}\mathbf{h}_{3(m)}\mathbf{\Pi}_2 - \ldots - u_{t-S}\mathbf{h}_{3(m)}\mathbf{\Pi}_S =$$
$$= u_t\mathbf{h}_{3(m)} - u_{t-1}\mathbf{h}_{3(m)} + u_{t-1}\mathbf{h}_{3(m)} - u_{t-1}\mathbf{h}_{3(m)}\mathbf{\Pi}_1 - u_{t-2}\mathbf{h}_{3(m)}\mathbf{\Pi}_2 + u_{t-1}\mathbf{h}_{3(m)}\mathbf{\Pi}_2 - u_{t-1}\mathbf{h}_{3(m)}\mathbf{\Pi}_2 +$$

$$- u_{t-3}\mathbf{h}_{3(m)}\mathbf{\Pi}_3 - \ldots - u_{t-S}\mathbf{h}_{3(m)}\mathbf{\Pi}_S = \Delta u_t\mathbf{h}_{3(m)} + u_{t-1}\mathbf{h}_{3(m)}(\mathbf{I} - \mathbf{\Pi}_1 - \mathbf{\Pi}_2) + \Delta u_{t-1}\mathbf{h}_{3(m)}\mathbf{\Pi}_2 +$$

$$- u_{t-3}\mathbf{h}_{3(m)}\mathbf{\Pi}_3 - \ldots - u_{t-S}\mathbf{h}_{3(m)}\mathbf{\Pi}_S = u_{t-1}\mathbf{h}_{3(m)}(\mathbf{I} - \mathbf{\Pi}_1 - \mathbf{\Pi}_2 - \ldots - \mathbf{\Pi}_\mathbf{S}) + \Delta u_t\mathbf{h}_{3(m)} + \Delta u_{t-1}\mathbf{h}_{3(m)}\mathbf{\Pi}_2 +$$

$$+ u_{t-1}\mathbf{h}_{3(m)}(\mathbf{\Pi}_3 + \ldots + \mathbf{\Pi}_\mathbf{S}) - u_{t-3}\mathbf{h}_{3(m)}\mathbf{\Pi}_3 - \ldots - u_{t-S}\mathbf{h}_{3(m)}\mathbf{\Pi}_S.$$

Successively adding and subtracting components $u_{t-2}\boldsymbol{\mu}_3(\mathbf{\Pi}_3 + \ldots + \mathbf{\Pi}_\mathbf{S})$, $u_{t-3}\boldsymbol{\mu}_3(\mathbf{\Pi}_4 + \ldots + \mathbf{\Pi}_\mathbf{S})$, … , $u_{t-S+1}\boldsymbol{\mu}_3\mathbf{\Pi}_S$, we can ultimately write:

$$u_t\mathbf{h}_{3(m)}\mathbf{\Pi}(L) = -u_{t-1}\mathbf{h}_{3(m)}\mathbf{B}\mathbf{A}^T + \Delta u_t\mathbf{h}_{3(m)} + \sum_{s=1}^{S-1}\Delta u_{t-s}\mathbf{h}_{3(m)}(-\mathbf{\Gamma}_s) \tag{3.26}$$

By performing transformations analogous to these above, we can present the last component as:

$$\mathbf{h}_{4(m)}\mathbf{\Pi}(L) = -(t-1)u_{t-1}\mathbf{h}_{4(m)}\mathbf{B}\mathbf{A}^T + [tu_t - (t-1)u_{t-1}]\mathbf{h}_{4(m)} +$$

$$+\sum_{s=1}^{S-1}[(t-s)u_{t-s} - (t-s-1)u_{t-s-1}]\mathbf{h}_{4(m)}(-\mathbf{\Gamma}_s) =$$

$$= -(t-1)u_{t-1}\mathbf{h}_{4(m)}\mathbf{B}\mathbf{A}^T + [t\Delta u_t + u_{t-1}]\mathbf{h}_{4(m)} + \sum_{s=1}^{S-1}[(t-s)\Delta u_{t-s} + u_{t-s-1}]\mathbf{h}_{4(m)}(-\Gamma_s).$$

(3.27)

After substituting the formulae (3.23)-(3.27) in (3.22), we arrive at VEqCM representation for the model with a constant, a trend, and adjustments of the constant and the parameter associated with the trend (see a less complex case considered by P. Saikkonen, H. Lütkepohl, C. Trenkler [2004]):

$$\Delta\mathbf{v}_{(m)t} = [\mathbf{v}_{(m)t-1}\mathbf{B} + \mathbf{g}_{1(r)} + (t-1)\mathbf{g}_{2(r)} + u_{t-1}\mathbf{g}_{3(r)} + (t-1)u_{t-1}\mathbf{g}_{4(r)}]\mathbf{A}^{\mathrm{T}} +$$

$$+\sum_{s=1}^{S-1}\Delta\mathbf{v}_{(m)t-s}\mathbf{\Gamma}_s + \mathbf{f}_{2(m)} + \sum_{s=0}^{S-1}\Delta u_{t-s}\mathbf{f}^s_{3(m)} + \sum_{s=0}^{S-1}[(t-s)\Delta u_{t-s} + u_{t-s-1}]\mathbf{f}^s_{4(m)} + \xi_{(m)t}$$

(3.28)

for $t = S+1, S+2, \ldots$
where:

$$\mathbf{g}_{1(r)} = -\mathbf{h}_{1(m)}\mathbf{B};\quad \mathbf{g}_{2(m)} = -\mathbf{h}_{2(m)}\mathbf{B};\quad \mathbf{g}_{3(m)} = -\mathbf{h}_{3(m)}\mathbf{B};\quad \mathbf{g}_{4(r)} = -\mathbf{h}_{4(m)}\mathbf{B};$$

$$\mathbf{f}_{2(m)} = \mathbf{h}_{2(m)} - \mathbf{h}_{2(m)}\sum_{s=1}^{S-1}\mathbf{\Gamma}_s = \mathbf{h}_{2(m)}\mathbf{\Psi},$$

$$\mathbf{f}^s_{3(m)} = \begin{cases}\mathbf{h}_{3(m)} & \text{dla}\quad s=0\\ -\mathbf{h}_{3(m)}\mathbf{\Gamma}_s & \text{dla}\quad s=1,\ldots,S-1\end{cases}$$

$$\mathbf{f}^s_{4(m)} = \begin{cases}\mathbf{h}_{4(m)} & \text{dla}\quad s=0\\ -\mathbf{h}_{4(m)}\mathbf{\Gamma}_s & \text{dla}\quad s=1,\ldots,S-1\end{cases}.$$

By defining the matrices:

$$\mathbf{G} = \begin{bmatrix} \mathbf{g}_{1(r)} \\ \mathbf{g}_{2(r)} \\ \mathbf{g}_{3(r)} \\ \mathbf{g}_{4(r)} \end{bmatrix}, \quad \mathbf{d}_{(z)t-1} = \begin{bmatrix} 1 \\ t-1 \\ u_{t-1} \\ (t-1)u_{t-1} \end{bmatrix}^T, \quad \text{and} \quad \Delta\mathbf{d}_{(z)t-s} = \begin{bmatrix} 0 \\ 1 \\ \Delta u_{t-s} \\ (t-s)\Delta u_{t-s} + u_{t-s-1} \end{bmatrix}^T,$$

$\mathbf{F}_s = \begin{bmatrix} \mathbf{f}_{1(m)} \\ \mathbf{f}_{2(m)} \\ \mathbf{f}_{3(m)} \\ \mathbf{f}_{4(m)} \end{bmatrix}$, where $\mathbf{f}_{1(m)}$ is any 1xM vector, we can write (3.28) in the following form:

$$\Delta\mathbf{v}_{(m)t} = \mathbf{v}^*_{(n)t-1}\mathbf{B}\mathbf{A}^T + \sum_{s=1}^{S-1}\Delta\mathbf{v}_{(m)t-s}\mathbf{\Gamma}_s + \sum_{s=0}^{S-1}\Delta\mathbf{d}_{(z)t-s}\mathbf{F}_s + \xi_{(m)t} \tag{3.29}$$

where:

$\mathbf{v}^*_{(n)t-1} = [\mathbf{v}_{(m)t-1} \;\vdots\; \mathbf{d}_{(z)t-1}]$ - the 1xN vector of stochastic and deterministic variables, $N = M + Z$,

$\mathbf{B}^* = \begin{bmatrix} \mathbf{B} \\ \dots \\ \mathbf{G} \end{bmatrix}$ – the NxR matrix of cointegrating vectors,

$\mathbf{G}$ - the ZxR matrix of parameters associated with the deterministic variables in cointegrating relations,

$\mathbf{F}_s$ - the ZxM matrices of parameters associated with the deterministic variables outside the cointegration space,

$n = 1, \dots, N$.

The Johansen method allows decomposing the matrix $\mathbf{\Pi}^* = \mathbf{B}^*\mathbf{A}^T$ and finding weights and mutually orthogonal cointegrating vectors that define the cointegration space. Every convex linear combination of the cointegrating vectors also belongs to this space and represents a cointegrating vector. Hence, there can exist an infinite number of cointegrating vectors. In most cases, the lineary independent cointegrating vectors do not have economic interpretation, thus the evaluation of the obtained results becomes much more difficult. This problem is solved by decomposing the matrix $\mathbf{\Pi}^*$ in such a way so that the resulting cointegrating vectors will correspond to (long-term) structural relationships satisfying economic theory. At least R^2 restrictions have to be imposed in order to identify the long-term structure. The first group of restrictions arises from the normalisation of particular cointegrating vectors with respect to successive variables. The second group consists of zero restrictions causing that the appropriate variables are omitted from the long-term relations (exclusion restrictions). The third group of restrictions is imposed under the assumed economic theory. In most cases, these are the homogeneity conditions that have to be met by some cointegrating vectors representing the long-term relationships and cross-equation restrictions occurring between the parameters of various cointegrating vectors. After allowing for the normalizing restrictions, at least $R(R-1)$ restrictions are necessary (under the dimension condition) to identify the matrix $\mathbf{B}^*$. When parameters are estimated using the FIML, it is irrelevant how many restrictions are imposed in each of the equations.

The system is underidentified if the number of restrictions V meets the condition $V < R^2$, but for $V > R^2$ it is overidentified. In the latter case, the restrictions can be tested by means of a standard likelihood ratio test.

The weights, i.e. matrix $\mathbf{A}$ parameters, can also be restricted. When all elements in the m-th line are zeros, the selected m-th variable is weakly exogenous in the long-run (there are no adjustments to the long-run trajectories resulting from the cointegrating relations). However, when all parameters in the appropriate column of matrix $\mathbf{A}$ are zero, then a given long-run relation does not affect the formation of any of the system variables. This situation should be viewed as quite pathological, because its consequences are a reduced order of matrix $\mathbf{A}$, and thus of matrix $\mathbf{B}$ and doubts thrown on the already identified dimension of the cointegration space.

Distinguishing the last $M-J$ in the vector M of the system stochastic variables, we can write (see: I. Harbo et al. [1998]):

$$[\Delta\mathbf{y}_{(j)t}\ \Delta\mathbf{x}_{(k)t}] = \mathbf{y}^*_{(n)t-1}\mathbf{B}^*[\mathbf{A}_1^T\ \mathbf{A}_2^T] + \sum_{s=1}^{S-1}\Delta\mathbf{y}_{(m)t-s}[\mathbf{\Gamma}_{1s}\ \mathbf{\Gamma}_{2s}] + \mathbf{d}_{(z)t}[\mathbf{F}_1\ \mathbf{F}_2] + [\xi_{1(j)t}\ \xi_{2(k)t}] \tag{3.30}$$

where x_{kt} denote variables with indices ranging from $J+1$ to M and $\mathbf{A}^T = [\mathbf{A}_1^T\ \mathbf{A}_2^T]$, $\mathbf{\Gamma}_s = [\mathbf{\Gamma}_{1s}, \mathbf{\Gamma}_{2s}]$, $\mathbf{F} = [\mathbf{F}_1\ \mathbf{F}_2]$, $\xi_{(m)t} = [\xi_{1(j)t}\ \xi_{2(k)t}]$, $j = 1,\ldots,J$, $k = J+1,\ldots,M$. By making the following partitioning of the variance-covariance matrix of the random terms in the original model $\mathbf{\Omega} = \begin{pmatrix} \mathbf{\Omega}_{11} & \mathbf{\Omega}_{12} \\ \mathbf{\Omega}_{21} & \mathbf{\Omega}_{22} \end{pmatrix}$ and transforming the formula we obtain:

$$\Delta\mathbf{y}_{(j)t} = \mathbf{y}^*_{(n)t-1}\mathbf{B}^*[\mathbf{A}_1^T - \mathbf{A}_2^T\Upsilon] + \sum_{s=1}^{S-1}\Delta\mathbf{y}_{(m)t-s}[\mathbf{\Gamma}_{1s} - \mathbf{\Gamma}_{2s}\Upsilon] + \Delta\mathbf{x}_{(k)t}\Upsilon + \mathbf{d}_{(z)t}[\mathbf{F}_1 - \mathbf{F}_2\Upsilon] + \hat{\xi}_{(j)t} \tag{3.31a}$$

$$\Delta\mathbf{x}_{(k)t} = \mathbf{y}^*_{(n)t-1}\mathbf{B}^*\mathbf{A}_2^T + \sum_{s=1}^{S-1}\Delta\mathbf{y}_{(m)t-s}\mathbf{\Gamma}_{2s} + \mathbf{d}_{(z)t}\mathbf{F}_2 + \xi_{(k)t} \tag{3.31b}$$

where:

$\Upsilon = \mathbf{\Omega}_{22}^{-1}\mathbf{\Omega}_{21}$,

$\hat{\xi}_{(j)t} = [\xi_{1(j)t} - \xi_{2(k)t}\Upsilon]$ and is independent of $\xi_{(k)j}$.

The equations (3.31a) and (3.31b) are called a conditional model and a marginal model, respectively. If $\mathbf{A}_2^T = \mathbf{0}$, then the variables x_{kt} are weakly exogenous and estimation of the conditional model parameters generates effective estimators of matrix $\mathbf{B}^*$ parameters. Additionally, the omission of (3.31b) in the finite and usually small samples offers significantly better results.

It may happen that an appropriate column, e.g. the m-th column of all matrices $\mathbf{\Gamma}_s$ is all zeros, which combined with weak exogeneity means that the m-th variable evolves independently of both long-run trajectories of equilibrium determined by the cointegrating vectors and short-run impacts represented by $\Delta y_{m,t-s}$.

It is asymptotically irrelevant, whether the identifying restrictions are tested before or after the dynamic, short-term model structure has been established. Still, in the case of relatively small samples that are most typical in modelling practice, the tests rapidly lose their power, if the dynamic structure has not been determined beforehand. On the other hand, the distributions of statistics determining the order of cointegration order are not known, if restrictions have been imposed on the structure of lags. Additionally, the power of the tests drops rapidly, if the errors are not normally distributed, or if autocorrelated. The results of the Monte Carlo experiments also indicate that the order of cointegration should be (re)examined after the model has been marginalised. Hence, a correct modelling strategy comprises the following, sequentially arranged steps (see: J.V.Greenslade, S.G.Hall, S.G.B.Henry [1999]).

First, the VAR model parameters are estimated and the order of lags is established. The order of lags should guarantee that the hypothesis of normal distribution of the errors will not be rejected under a conventionally accepted test size.

At the second stage, the order of cointegration is determined, using a classical Johansen procedure.

At the third stage, the model is marginalised. Because the critical values of the test for the size of the cointegration space depend on the number of weakly exogenous variables, this procedure has an iterative character: after the number of the cointegrating vectors has been established, the selected variable is tested for weak exogeneity. If the null hypothesis is not rejected, the variable is added to the set of the weakly exogenous variables and the number of the cointegrating vectors is redefined. A weakly exogenous variable means that none of the cointegrating relations is important for explaining its variance. It is reasonable to expect that for a properly assembled set of variables, all variables recommended under economic theory should prove endogenous. If otherwise, an important variable was omitted from the system or the formulated economic hypotheses were not true. In the case of small-size samples, the hypothesis of weak exogeneity of a variable can be rejected even though its status does not stir any doubts (for instance, the index of world prices is certainly weakly exogenous in the economic model of a country having a very small share in international trade). When the model cannot be re-specified, the weak exogeneity of this variable needs to be assumed. Unfortunately, this affects the distribution of statistics determining the dimension of the cointegration space and the critical values. This approach may therefore lead to wrong conclusions.

The list of the exogenous variables complete, at the fourth stage at least R^2 restrictions are imposed on the long-run structure of the model to guarantee the identification of the matrices $\mathbf{B}^*$ and $\mathbf{A}_2$ and all model parameters are estimated. If the number of restrictions exceeds R^2, then the likelihood ratio test is applied to verify the correctness of the set of restrictions.

It is possible to impose additional restrictions and then test the new set of restrictions by estimating the model parameters. As a result, a full set of restrictions postulated by economic theory that were not rejected in the course of empirical verification is generated.

The fifth step consists in determining the dynamic structure of the model (i.e. elements of the matrix $\mathbf{\Gamma}_s$). Because at this stage of the procedure the whole model is based on the stationary variables, the significance of particular lags can be verified using conventional statistics.

Cointegration makes it possible to build a bridge between statistical inference and a concept of system equilibrium and this fact should be recognized as the source of its huge popularity. On one hand, the application of the VEqCM (SVEqCM) models allows making the most of economic theory (by imposing restrictions on the long-run parameters) and on the other hand, the common stochastic trends can be eliminated, which is a prerequisite for avoiding the danger of spurious regressions.

References:

Bärdsen G., Eitriheim 0, Jansen E.S., Nymeoen R., 2005, *The Econometrics of Macroeconometric Modelling,* Oxford University Press, Oxford.

Bodkin R.Cz., Klein L.R., Marwah K. (ed.), 1991, A *History of Macroeconometric Model Building,* B. Blegar, Aldershot.

Garratt A., Lee K., Pesaran M.H., Shin Y., 2006, *Global and National Macroeconometric Modeling: A Long-Run Structural Approach,* Oxford University Press, Oxford.

Greenslade J.V., Hall S.G., Henry S.G.B. [1999], *On the Identification of Cointegrated Systems in Small Samples: Practical Procedures with an Application to UK Wages and Prices,* Centre for Economic Forecasting London Business School, Discussion Paper.

Grzęda-Latocha R., 2005, *Ekonometryczna analiza koniunktury gospodarczej w krajach sfery euro,* (*Econometric Analysis of Business Cycle In the EUROLAND*),

„Ekonomista", vol. 5, pp. 621-640.
Harbo, I., Johansen, S., Nielsen, B., Rahbek, A., 1998, *Asymptotic Inference on Cointegrating Rank in Partial Systems.* Journal of Business and Economic Statistics vol. 16, pp. 388-399.
Jorgenson D.W., 2000, *Information Technology and the US Economy,* „American Economic Review", vol. 91, pp. 1-32.
Jorgenson D.W., Ho M.S., Strip K.J., 2002, *Projecting Productivity Growth. Lessons from the Resurgence, Federal Reserve Bank of Atlanta,* „Economic Review", vol. 87, pp. 1-14.
Klein L.R., Welfe A., Welfe W., 1999, *Principles of Macroeconometric Modelling,* North Holland, Amsterdam.
Saikkonen P., Lütkepohl H., Trenkler C., 2004, *Break date estimation and cointegration testing in VAR processes with level shift*, European University Institute, Working Paper No. 2004/21.
Solow R., 1957, *Technical Change and Aggreggate Production Function,* „Review of Economics and Statistics", vol. 39, pp. 312-320.
Welfe A., 2009, *Ekonometria. Metody i ich zastosowanie* (*Econometrics. Methods and Applications*), IV edition, PWE, Warsaw.
Welfe W., 1992, *Ekonometryczne modele gospodarki narodowej Polski,* (*Econometric Models of the Polish Economy*), PWE, Warsaw.
Welfe W., 2000, *Empiryczne modele wzrostu gospodarczego* (*Empirical Models of Economic Growth*), „Ekonomista", No. 4., pp. 483-499.
Welfe W., 2008, *A Knowledge-Based Economy: New Directions of Macromodelling*, "International Advances in Economic Research", vol. 14, pp. 167-180.
Welfe W. (ed.), 2001, *Ekonometryczny model wzrostu gospodarczego,* (*Econometric Model of Economic Growth*), Wydawnictwo Uniwersytetu Łódzkiego, Łódź.
Welfe W. (ed.), 2004, *Długookresowy, makroekonometryczny model W8D-2002 gospodarki polskiej,* (*Long-term Macroeconometric Model of the Polish Economy W8D-2007*), „Acta UŁ Folia Oeconomica", vol. 172, Wydawnictwo Uniwersytetu Łódzkiego, Łódź.
Welfe W., 2005, *Stylized Empirical Model of Economic Growth*), in: Welfe A., Welfe W. (ed.), *Macromodels 2004. Problems of building and estimation of econometric models*, . „Acta UŁ Folia Oeconomica", vol. 190, Wydawnictwo Uniwersytetu Łódzkiego, Łódź, pp. 109-125.
Welfe W., Welfe A., 2004, *Ekonometria stosowana,* (*Applied Econometrics*), II edition, PWE, Warsaw.
Whitley J.D. 1994, *Course in Macroeconomic Modelling and Forecasting,* Harvester/Wheatsheaf, New York.
Zienkowski L. (ed.), 2003, *Wiedza a wzrost gospodarczy,* (*Knowledge and Economic Growth*), Wydawnictwo Naukowe SCHOLAR, Warsaw.

Chapter 4

Total factor productivity. The embodied capital of knowledge

4.1 Introduction

The effects of technological changes we observe manifest themselves not only through the production capacity of economies, the structure and organization of production, the productivity of labour, the capital-labour ratio, but also, or perhaps above all, through the living standard of societies. Progress is inseparable from reaching higher levels of quality in the sphere of production, but also in the sphere of knowledge, awareness, skills and qualifications. Technological progress sets a given economy on the path of development and *ipso facto* guarantees the possibility of meeting individual and public needs in a society. The advancing technology that becomes increasingly effective while placing stronger emphasis on protecting the natural environment, saving the resources and improving the quality of the manufactured goods can be viewed as the major source of progress and economic growth.

Technological progress results in the growth of production stream that does not involve higher production inputs. Technological progress is represented not only by changes in the technology of production, but also by changes in the organization of production that aim at improving the efficiency of management. Therefore, when discussing the issue of technological development, we very often mean new technological and organizational solutions, because technological changes frequently affect the organization of production, and separating these aspects factors seems impossible.

4.2 The macroeconomic definition of technological progress

The term 'technological progress' is frequently applied exclusively to changes in a technique or a technology of production, and reflects the advancement towards higher quantitative and qualitative levels of technology, generating positive economic, social and environmental effects vis-à-vis the starting point (Białoń, Obrębski, [1989], p.19). Therefore, this term has a strong connection with the production process and more precisely with these changes in the production structure and in the inputs of the production factors that make the production process more efficient. However, the technological progress affects not only the technical side of production, but also its organizational part. A broader definition of technological progress covering a larger number of aspects can be found in the Solow's work of 1967. Solow analysed the effects of technological progress with respect to the neoclassical production function, which is a basic tool of the neoclassical models of economic growth. His definition of economic growth (Solow, [1967], p. 48) says that when technological progress is considered in abstractive terms, which approach is specific to economists, it is quite natural for us to imagine a typical production graph having production inputs measured along the axes and a group of curves of conventionally-shaped identical products, and to claim that when technological progress takes place these curves move in a way that a given expenditure can fund the manufacture of more products, or the same product can be delivered at a lower cost. Hence, every increase in the stream of production that is unrelated to the accumulation of real capital and labour arises from technological progress.

The Solow's concept of technological progress builds on a neoclassical two-factor production function[1] having the following form:

$$Y = F(K, L) \tag{4.1}$$

where:

Y – volume of output;

K – capital stock;

L – labour force.

In principle, two basic kinds of technological progress can be distinguished, namely the disembodied and embodied progress (Allen, [1975], p. 235). The first covers all the stock of the currently applied production factors, treating them as homogenous. As Allen puts it, this type of technological progress represents technical "know-how" and "falls like a manna from heaven". The disembodied technological progress makes production grow because of rising effectiveness of all available stock of production factors, but the internal structure remains unchanged (Juszczak-Szumacher, [1996], pp. 63-64). This type of progress is analysed with respect to neoclassical theory of production.

The embodied technological progress does not affect all available production factors, but only some groups of capital goods – usually the machinery manufactured and installed on a current basis and its operation (Allen, [1975], p. 236). This means that the embodied technological progress stimulates output growth owing to higher effectiveness of factor inputs, which involved changes in their internal structure. Therefore, the neoclassical assumption of the homogeneity of production inputs is cancelled, and the inputs are rather viewed as a group of different "vintages". The latest "vintages" are more efficient and thus they carry technological progress with them (Juszczak-Szumacher, [1996], p. 63).

Using a production function as an analytic method for investigating technological progress implies treating the latter as another production factor. Assuming that this progress is exogenous, it can be added to the production function as a time variable t or a time function. The effects of technological

[1] A neoclassical production function is characterized by positive and decreasing marginal productivity of each production factor, it is homogenous of degree one and fulfils the Inada's conditions ([Barro, Sala-i-Martin, [1995], p. 16-17). Cf. also chapter two.

progress can be also analysed in terms of changing effectiveness of particular factors. The changes may affect a single production factor as well as their entire group. As a result, we can talk about a technological progress improving the effectiveness of capital (capital augmenting) or labour (labour augmenting), or of both of the factors at the same time (factor augmenting). In order to allow for each of these tendencies, a given factor (or both) has to be expressed as units of effectiveness. The production function with inputs measured with respect to the units of effectiveness takes the following form:

$$Y = F[a(t)K, b(t)L] \tag{4.2}$$

where

$a(t), b(t)$ are such time functions that $a(t) \geq 1$, $b(t) \geq 1$ and $\frac{da}{dt} \geq 0$, $\frac{db}{dt} \geq 0$.

According to expressions $a(t)K$ and $b(t)L$, the inputs of capital and labour are not measured in natural units, but in the units of effectiveness. This means that under constant inputs of factors production may expand because of improving effectiveness of the given production factor (factors)[2].

It should to be added that the tendencies characterising technological progress induced by increasing effectiveness of the production factors are not related to the notion of embodied technological progress. Higher effectiveness of a given factor (or both the factors together) is showing the entire available stock of this factor (factors), without changing its (their) internal structure. The homogeneity assumption of the production factors concerns factors measured in natural units and in units of effectiveness.

According to the previously adopted definition, the effects of technological progress are equated with a shift of the production allowing a given production volume to be obtained using a smaller amount of inputs (or larger production can be derived from the same amount of inputs).

However, the shift of the production function can be accompanied by changing shares of the production factors, which in the case of a two-factor function means different proportions of capital and labour. A shift in the production function can also occur without any changes in the capital-to-labour ratio $\left(K/L\right)$. Then technological progress is neutral. When the ratio increases together with the shifting production function, then the technological progress is labour-saving

[2] Thus defined technological progress increases the efficiency of both production factors at the same time. However, it can increase the efficiency of only one factor of production. Therefore, this progress can be said to be capital augmenting when $Y = F(a(t)K, L)$, or labour augmenting when $Y = F(K, b(t)L)$.

(capital consuming). When the ratio declines, the technological progress is classified as labour consuming (capital saving).

The neutrality of technological progress is related to stability of relations. This typically concerns the aforementioned capital-labour ratio (neutral technological progress as defined by Hicks), but stability of the capital-output ratio (Harrodian neutral technological progress), stability of labour productivity (Solow), or stability of the relations between the rates of growth of the capital-output ratio and the growth rate of labour productivity (Kalecki, [1963]) are also postulated. A discussion of the various types of neutrality of technological progress goes beyond the topic of this chapter. It seems, however, that the further deliberations make it justifiable to define the neutrality of technological progress as defined by Hicks. He claims that technological progress is neutral if the marginal rate of substitution of labour with capital does not change under a constant capital-labour ratio. Therefore, it is characterised by symmetry, meaning that it equally improves the effectiveness of both production factors. According to Hicks, the neutral technological progress reveals the factor-augmenting propensity under an additional assumption that $\frac{da}{dt} = \frac{db}{dt} = \alpha(t)$, with $\alpha(t) > 1$ and $\frac{d\alpha}{dt} > 0$. As a result, Hicks' two-factor production function with the neutral technological progress has the form (Barro, Sala-i-Martin, [1995], p. 33):

$$Y = A(t) \cdot F(K, L), \tag{4.3}$$

where $A(t)$ – factor describing technology of production, and $\frac{dA}{dt} \geq 0$.

4.3 The role of product, process and organizational innovations in generating technological progress

The concept of technological progress is combined with the concept of innovation. The latter means an act of introducing a qualitative change in the economy by starting the manufacturing of a new product (application of a new process), as well as the product (process) itself (Gomułka, [1998], p. 17), whereas the process of innovation is represented by all research, development and implementation activities undertaken at a given time within the analyzed sphere that aim at practical application of better production methods and means of production, as well as manufacturing of new or improved products (Krajewski, [1985], p. 20). The theory of innovation is thought to have been formulated by Schumpeter[3]. He was one of the first to define the particular role played by innovations in economic growth, claiming that they were one of the main factors,

[3] We can also mention Babbege and Rae, the 19th century economists, who are considered the forgotten precursors of Schumpeter's theory of innovation (Mikosik, [1993], p. 68).

in addition to enterprise and credit, which determine economic growth. (Mikosik, [1993], pp. 67-74; Blaug, [1994], pp. 446-447; Tokarski, [2001a], p. 13). He also introduced to economic theory terms such as imitation, innovation and invention (Kasperkiewicz, [1989], pp. 13-14).

Innovations are typically divided into those affecting products and processes. Product innovations involve the introduction of new or improvement of existing products, whereas the process-related innovations introduce new, improved techniques of production. This division, however clear from the theoretical point of view, is in fact a little artificial, and the empirical distinction between effects generated by the introduction of product innovations and process innovations is often impossible. It is so, because product innovations intertwine with process innovations; the new, improved products require the launching of new technical equipment (Krajewski, [1985], pp. 23-29; Świtalski, [2005], pp. 87-105). It is also difficult to establish whether a process innovation results from the application of a product innovation (such as new machinery, equipment), or perhaps from the modification of the production processes (i.e. a more effective use of the productivity factors, Świtalski, [2005], p. 73).

The concept of innovation may also concern to the organizational aspect of production. All changes in a production process that result from introduction of new methods to the organization and management of the production process and marketing can be termed as an organization innovation (Krajewski [1985], pp. 23-29; Gomułka, [1998]; pp. 17-20).

There is one more type of innovation, namely the social innovation, which is indirectly connected with organizational innovations. It allows persons having different skills and knowledge to co-operate efficiently within an organization (Drucker, [1992], p. 40).

Innovation processes in the economy result from factors existing on both the supply side and demand side of the economy. Two main concepts describing the way innovations emerge and function can be distinguished (Nordhaus, [1976], pp. 67-70). One is the so-called supply approach (pioneered by Schumpeter). It is based on the assumption that each economy has a sector dedicated to creating new techniques and products (otherwise R&D), and then spreading the effects of its work among other sectors of economy. The only source of inventions created in an economy is creative aspirations of their designers and the sole purpose of their creation is their being patented at some later point of time. They usually go through a long and costly process of improvements, their implementation takes time and the creators are usually left outside. When an invention is finally applied, then innovation occurs. The concept of demand is poles apart with the supply concept. It assumes that new, upgraded products is mainly created owing to the demand of potential buyers (companies, local authorities, households), and that they are introduced promptly, with active involvement of their creators. The results of empirical research show that the demand concept of innovations clearly prevails in the economy (Kasperkiewicz, [1989]; pp. 65-75).

When effects of technological progress are being analysed in terms of the production function, we tend to concentrate on effects of process innovations. Distinguishing the process innovations from product innovations is not easy, especially when such an analysis is carried out at the macro economic level. Innovations related to production processes and increasing their effectiveness frequently result from improvements and better quality of the relevant machines and equipment - thus they are connected with product innovations. The role of the organizational innovations should not be omitted, either, as these can significantly raise the effectiveness of the production processes.

4.4 The macroeconomic measures of the effects of technological progress

In micro economic terms, the effects of technological progress are typically assessed based on individual innovative projects (Juszczak-Szumacher, [1996]; p. 58). In contrast, the macroeconomic approach usually equates the effects of technological progress with changing effectiveness of larger economic structures. The changes are usually expressed as alterations in:

a) the structure of production factors and in the relations among them,
b) the structure of production,
c) the productivity (or absorbability) of the production factors; these changes may affect only one factor (changes in the productivity of labour and/or capital, or changes in the capital or labour intensity) or a whole set of factors (changes in total factor productivity).

Among the measures placed in the first group, the capital-labour ratio is mentioned the most often. The effects of technological progress can manifest themselves through a higher capital-labour ratio that grows because human labour is being replaced with machinery. (Pawłowski, [1976], pp. 98-109). The increase in the capital-labour ratio can also be followed using the indicator of the mechanization of labour (defined as the share of the labour input needed by the mechanized work needs in total labour input; Król, [1974], pp. 29-30), as well as the increase in unit consumption of electric power. It is postulated to use the fixed assets renewal indicator that describes the share of investments project put in the total stock of capital. (Welfe, [1983]. pp.114-119).

Output changes can also reflect the effects of technological progress. The economy's total output is the sum of outputs produced by its individual sectors (sub-aggregates). Naturally, the level of disaggregation may vary and it exclusively depends on the specific goals of analysis. Structural changes should be understood as varying shares of particular sub-aggregates in relation to each other another and to the entirety (Stanek, [1990], p.19). The structural changes can be analyzed by means of technical coefficients (Czerwinski, [1982], pp. 257-274; Tomaszewicz, [1994], pp. 58-63) describing technical linkages between particular sectors (sub-aggregates). Changes in these coefficients indicate modifications taking place in the technical structure of the entire economy, as they reflect input changes in relation to output; under unit inputs declining in time are interpreted as the occurrence of technological progress (Stanek, [1990], p. 19).

In most cases, however, the effects of technological progress are treated as changes in the factor-intensity of production (i.e. absorbability of the production factors), or in the productivity of production factors. Absorbability (n_i) of the *i*-th production factor X_i is defined as the amount of the *i*-th factor per unit of output (*Y*):

$$n_i = \frac{X_i}{Y}, \quad \text{for } i \in \{1,2,\dots,n\} \tag{4.4}$$

If for all $i \in \{1,2,\dots,n\}$ relations:

$$\frac{d}{dt}\left(\frac{X_i}{Y}\right) < 0 \tag{4.5a}$$

or

$$\frac{dX_i}{dt}\cdot\frac{1}{X_i} < \frac{dY}{dt}\cdot\frac{1}{Y} \tag{4.5b}$$

are satisfied, then we can find technological progress in the growth of product Y. It is so, because equation (4.5a) or (4.5b) provides that the absorbability of each production factor decline in time[4]. This means that a given volume of production can be achieved using smaller inputs, which meets Solow's definition of technological progress. Assessment of the effects of technological progress based on the absorbability ratios can be found, for instance, in the works by Felbur ([1993], pp. 22-27 and next).

Factor productivity is inversely related to absorbability. It is the volume of production per unit of a given factor. Productivity (as well as absorbability) may refer to a single production factor, as well as a whole set of factors (total factor productivity, more in next sections of this chapter). The productivity of a given

[4] Technological progress occurs, when the relation (4.5) is met for all factors of production. Sometimes it is assumed that the relation (4.5) is met for some factors and the relation $\frac{d}{dt}\left(\frac{X_i}{Y}\right) = 0$ for others. It cannot be ascertained, however, whether technological progress is involved in the process of production growth, if the absorptiveness of certain production factors is on the rise (Stanek, [1990], pp. 16-18).

factor(s) is understood as the amount of production per unit of this factor(s). Productivity growth is often regarded as improved effectiveness of production processes. It is not strange, therefore, that the factor productivity changes are among the most popular measures in procedures analysing the effects of broadly understood technological progress. According to Solow definition technological progress arises from growing factor productivity (the isoquant of the production function moves to the origin of the system of coordinates), which growth frequently co-occurs with variations in relations between the factors (i.e. the isoquant shifts along the production function which can be interpreted as a growth in the capital-labour ratio).

When the productivity of a single factor of production is discussed, we typically mean the productivity of labour or capital. Both types of productivity reflect the effectiveness of a given factor as measured by the amount of output per unit of a given factor, that is:

$$PROD_{it} = \frac{Y_t}{X_{it}}, \tag{4.6}$$

where:

Y_t – the magnitude of output in period t (usually given as physical units or constant price values),

X_{it} – magnitude of i-th production factor at time t (usually given as physical units or in constant prices).

If both output and inputs are assumed to be changing continuously, then after differentiating the relation (4.6) with respect to time and then dividing both its sides by $Prod_{it}$, we can rewrite the relation in terms of growth rates, i.e.:

$$\frac{dPROD_{it}}{dt} \cdot \frac{1}{PROD_{it}} = \frac{dY_t}{dt} \cdot \frac{1}{Y_t} - \frac{dX_{it}}{dt} \cdot \frac{1}{X_{it}} \tag{4.7}$$

The above means that the growth rate of given factor's productivity is the difference between the output growth rate and the growth rate of the given factor.

As a result of technological progress factor productivity rises. However, when technological progress is being analysed, we need to take into account the substitution of production factors, which is typically understood as replacing human labour by machinery and equipment. Therefore, its effect is a higher capital-labour ratio.

A comprehensive analysis of the effects of technological progress should address as many characteristics of the progress as possible. It is also possible to construct a single synthetic standard as a combination of partial standards by means of taxonomic methods measuring the distance between the analysed structure and some generally accepted model (Juszczak-Szumacher, [1996]). This approach can be found in the works by Białoń ([1974], pp. 210-220) and Kudrycka ([1990], pp. 6-13). However, its obvious shortcoming is the impossibility of finding the absolute value of this measure and thus analyses of the investigated structures in time are prevented. Still another option is the concept described in the work by Cieślak ([1976], pp. 99-103) that uses an indicator measuring certain absolute level of technology. The advantage of this approach is that it allows analyses of structures in both space and time.

An alternative to the aforementioned methods is analysis based on the production function. Theoretically, this approach may seem straightforward, but in empirical analyses many problems need to be solved, two of which seem to be crucial. Firstly, what should be the form of this function and (then) what estimation methods of its parameters are appropriate? Secondly, which factors of production should be taken into account? According to Żółtowska ([1997], p. 5), a considerable discrepancy exists between the analytic forms of the production function discussed in theoretical works and these operated in empirical research. This situation is due to the fact that the function's analytic forms belong to the class of non-linear functions and in most cases they do not have the linear transformations. Therefore, the iterative methods depend have to be applied to estimate the function parameters. Efficiency and precision of especially the iterative methods depend, *inter alia,* on the amount of statistical data, and the available data are usually in a short supply. Notwithstanding these rather significant limitations, many authors use the production function to assess the effects of technological progress (see also Tinbergen, [1959]; Solow, [1957]; Kim, Lau, [1994]; Welfe, [1992], [2000], [2001]; more on this topic can be found in the works by Welfe [2001] and Sztaudynger [2003]).

4.5 Total factor productivity and the methods of its measurement

Total factor productivity (TFP) corresponds to the volume of production per one unit of combined factors of production. Unlike the productivity of a single factor, the TFP growth is the main synthetic method used to measure changes in the effectiveness of production processes, brought about by broadly understood technological progress.

Two alternative methods can be applied to find total factor productivity. One takes advantage of the index method and the other uses the concept of production function. The index methods for calculating TFP are based on the concept of aggregated statistical indices (Laspeyres, Paasche, or Fisher indices), or on the concept of a continuous index (Divisia's index) and its discrete approximation (Törnquist index). TFP estimations using the production function usually consist in estimating the function's parameters, with the TFP level being accounted for in the production function as a function of time.

4.5.1. The index method of establishing TFP

Let us assume that a given economy manufactures n products using m production factors. According to the definition of TFP, its level is expressed as the volume of output per unit of combined factors. This approach is possible only in value terms. Consequently, the level of TFP is determined by the following relation:

$$TFP = \frac{p_1Y_1 + p_2Y_2 + \ldots p_nY_n}{q_1X_1 + q_2X_2 + \ldots q_mX_m} = \frac{\sum_{i=1}^{n} p_iY_i}{\sum_{j=1}^{m} q_jX_j} \tag{4.8}$$

where:

Y_i – the quantity of i-th manufactured product, $i \in \{1,2,\ldots n\}$;

p_i –price of one unit of i-th manufactured product, $i \in \{1,2,\ldots n\}$;

X_j –quantity of j-th production factor, $j \in \{1,2,\ldots m\}$;

q_j – price of one unit of j-th factor, $j \in \{1,2,\ldots m\}$.

To construct an aggregated TFP index comparing TFP levels for two different periods, t=0 and t=1, the values of inputs and outputs results must be converted into constant prices. The baseline used in comparisons can be prices derived from the initial period (t=0), or prices from the investigated period (t=1). In the first case, the TFP index is an aggregated statistical index as formulated by Laspeyres:

$$TFP^1 / TFP^0 = \left(\frac{\sum_{i=1}^{n} p_i^0 Y_i^1}{\sum_{j=1}^{m} q_j^0 X_j^1} \right) \div \left(\frac{\sum_{i=1}^{n} p_i^0 Y_i^0}{\sum_{j=1}^{m} q_j^0 X_j^0} \right) \tag{4.9}$$

where:

Y_i^0, Y_i^1 - the quantity of i-th product in the initial period (t=0) and in the analyzed period (t=1), respectively;

X_j^0, X_j^1- the quantity of j-th factor in the initial period (t=0) and in the analyzed period (t=1), respectively;

p_i^0, q_j^0- prices of one unit of product and of factor applied in the initial period (t=0).

If the product and factor prices are set at the level of prices from the analyzed period (t=1), then the index of TFP changes for the two periods is an aggregated statistical index as formulated by Paasche:

$$TFP^1 / TFP^0 = \left(\frac{\sum_{i=1}^{n} p_i^1 Y_i^1}{\sum_{j=1}^{m} q_j^1 X_j^1} \right) \div \left(\frac{\sum_{i=1}^{n} p_i^1 Y_i^0}{\sum_{j=1}^{m} q_j^1 X_j^0} \right) \tag{4.10}$$

Another option is applying the Fisher index, being the geometric mean of the Laspeyres and Paasche indices. This approach to estimating TFP can be found in the works by Kendrick and Grossman [1980].

To estimate TFP changes, Solow [1957], Griliches and Jorgenson [1967] utilized the concept of a continuous index developed by Divisia in the 1920s (Welfe, [1966], pp. 70-71). This concept builds on the assumption that a production process is continuous, and so the volume of output can be approximately treated as a continuous function of time. Accordingly, quantities of factors, their prices and TFP are continuous and differentiable functions of time. Consequently, relation (4.8) can be written as follows:

$$TFP(t) = \frac{\sum_{i=1}^{n} p_i(t)\, Y_i(t)}{\sum_{j=1}^{m} q_j(t)\, X_j(t)}. \tag{4.11}$$

Let $\sum_{i=1}^{n} p_i(t)\, Y_i(t) = \tilde{Y}(t)$ and $\sum q_j(t) X_j(t) = \tilde{X}(t)$. Then:

$$TFP(t) = \frac{\tilde{Y}(t)}{\tilde{X}(t)}. \tag{4.12}$$

Differentiating formula (4.12) with respect to time and dividing both its sides by $TFP(t)$ we obtain:

$$\frac{dTFP}{dt} \cdot \frac{1}{TFP(t)} = \frac{d\tilde{Y}(t)}{dt} \cdot \frac{1}{\tilde{Y}(t)} - \frac{d\tilde{X}(t)}{dt} \cdot \frac{1}{\tilde{X}(t)}. \tag{4.13}$$

At the same time:

$$\frac{d\tilde{Y}(t)}{dt} = \sum_{i=1}^{n} \frac{dp_i(t)}{dt} \cdot Y_i(t) + \sum_{i=1}^{n} p_i(t) \cdot \frac{dY_i(t)}{dt}. \tag{4.14}$$

Dividing both sides of relation (4.14) by $\tilde{Y}(t)$ and performing simple algebraic transformations, we arrive at:

$$\frac{d\tilde{Y}(t)}{dt}\cdot\frac{1}{\tilde{Y}(t)}=\sum_{i=1}^{n}w_i(t)\left(\frac{dp_i(t)}{dt}\cdot\frac{1}{p_i(t)}+\frac{dY_i(t)}{dt}\cdot\frac{1}{Y_i(t)}\right), \tag{4.15}$$

where $w_i(t)=\dfrac{p_i(t)Y_i(t)}{\sum_{i=1}^{n}p_i(t)Y_i(t)}$ is the share of the i-th product value in the total value of output.

In other words, the rate of growth of the aggregated output is a weighted sum of the growth rates of prices and quantities of manufactured goods with weights corresponding to the share (in value terms) of the given i-th type of manufactured goods in output.

Likewise, it can be shown that:

$$\frac{d\tilde{X}(t)}{dt}\cdot\frac{1}{\tilde{X}(t)}=\sum_{j=1}^{m}v_j(t)\left(\frac{dq_j(t)}{dt}\cdot\frac{1}{q_j(t)}+\frac{dX_j(t)}{dt}\cdot\frac{1}{X_j(t)}\right), \tag{4.16}$$

where $v_j(t)=\dfrac{q_j(t)X_j(t)}{\sum_{i=1}^{n}q_j(t)X_j(t)}$ stands for the share of i-th production factor in total production costs.

Given that prices of particular inputs and output remain unchanged in the analyzed period (they are expressed in constant prices), i.e.:

$$\sum_{i=1}^{n}\frac{dp_i(t)}{dt}\cdot\frac{1}{p_i(t)}=0 \quad\text{and}\quad \sum_{j=1}^{m}\frac{dq_j(t)}{dt}\cdot\frac{1}{q_j(t)}=0$$

we have:

$$\frac{dTFP}{dt}\cdot\frac{1}{TFP}=\sum_{i=1}^{n}w_t(t)\cdot\frac{dY_i(t)}{dt}\cdot\frac{1}{Y_i(t)}-\sum_{j=1}^{m}v_t(t)\cdot\frac{dX_j(t)}{dt}\cdot\frac{1}{X_j(t)}. \tag{4.17}$$

This means that the rate of TFP growth is a difference between the weighted sum of the growth rates of the volume of particular products manufactured in a given economy and the weighted sum of the growth rates of the volume of particular factors of production. The weights are the shares of particular products' quantities in total output (w_i) or in total costs (v_i).

Theoretically, the straightforward concept of the continuous index cannot be applied in empirical analyses. A discrete approximation of relation (4.17) is the

Törnqiust chain index (Gullickson, [1995], p. 19; Kunze, Jablonski, Klarquist, [1995], p. 4; Sudit, [1995]; p. 4420 written as:

$$\frac{TFP^1}{TFP^0}=\frac{\prod_{i=1}^{n}\left(\frac{Y_i^1}{Y_i^0}\right)^{\overline{w_i}}}{\prod_{j}^{m}\left(\frac{X_j^1}{X_j^0}\right)^{\overline{v_j}}}, \tag{4.18}$$

where $\overline{w_i}=\frac{w_i^0+w_i^1}{2}$ and $\overline{v_j}=\frac{v_j^0+v_j^1}{2}$.

The Divisia index, or more precisely its discrete approximation, is one of the more popular index methods applied to assess the rates of TFP growth[5] (Griliches and Jorgenson [1967]; Hulten, [1978]; Gullickson, [1995]; Wilder and Ziaul [2001]).

The presented index methods for examining TFP changes do not represent the whole set of available index methods used for this purpose. Nevertheless, they are the most popular in empirical investigations. Besides, the index methods can be applied to compare TFP levels in various economies. A case in point is the Malmquist index that helps compare TFP levels in two economic systems; the first to use it the context of productivity were Caves, Christiansen and Diwert in 1982. The starting point for constructing the Malmquist productivity index is finding the volume of output that the given economy can produce with the given stocks of factors, but using technology of production characteristic of another economy. By setting the actual volume of production in one economy against the volume of production that could be attained if the other economy's technology of production were applied, we can answer the question which of the technologies is more (or less) effective (Sudit, [1995], pp. 443-445; Maudos, Pastor and Serano, [1999], pp. 40–41).

4.5.2. Total factor productivity and the production function

Alternatively, we can try to find TFP by applying the concept of a production function. Taking first a simplifying assumption that the only factors are labour (L) and capital (K), the level of TFP (indicating the amount of output per unit of combined labour and capital) in period t is written as:

$$TFP_t=\frac{Y_t}{F(K_t,L_t)}. \tag{4.19}$$

[5] In many instances relation (4.18) is written as $\Delta\ln TFP_{1/0}=\sum_i \overline{w_i}\Delta\ln Y_{i\,1/0}-\sum_j \overline{v_j}\Delta\ln X_{j\,1/0}$ (Welfe, [2001], p. 120).

After transforming the expression (4.19) to be:

$$Y_t = TFP_t \cdot F(K_t, L_t) \tag{4.20}$$

and comparing it with (4.3), we see that the level of TFP corresponds to factor $A(t)$, which denotes the technology of production in a two-factor production function with neutral technological progress as defined by Hicks, (i.e. $TFP_t = A(t)$). Therefore, the level of TFP measures the distance by which the production function moves towards the origin of the system of coordinates (its axes correspond to capital K and labour L) as a result of higher effectiveness of these factors.

Differentiating both sides of (4.20) with respect to time, we have:

$$\frac{dY_t}{dt} = \frac{dTFP_t}{dt} \cdot F(K_t, L_t) + TFP_t \frac{\partial F}{\partial K} \cdot \frac{dK_t}{dt} + TFP_t \frac{\partial F}{\partial L} \cdot \frac{dL_t}{dt} \tag{4.21}$$

To simplify the formula, let us assume that $\frac{dY_t}{dt} = \dot{Y}_t$, $\frac{dTFP_t}{dt} = T\dot{F}P_t$, $\frac{dK_t}{dt} = \dot{K}_t$ $\frac{dL_t}{dt} = \dot{L}_t$.

Dividing relation (4.21) by Y, we arrive at:

$$\frac{\dot{Y}_t}{Y_t} = \frac{T\dot{F}P_t}{TFP_t} + \frac{\partial Y_t}{\partial K} \cdot \frac{K_t}{Y_t} \cdot \frac{\dot{K}_t}{K_t} + \frac{\partial Y_t}{\partial L} \cdot \frac{L_t}{Y_t} \cdot \frac{\dot{L}_t}{L_t} \,. \tag{4.22}$$

In other words, the rate of output growth (the left-hand side of relation (4.22)) is a sum of the rate of TFP growth (or of the neutral technological progress in Hicks's terms) and the weighted sum of the growth rates of primary factors whose weights equal to production elasticity towards respective factors.

If we assume, following Clark's marginal theory, that each factor of production is rewarded in relation to its marginal product:

$$\frac{\partial Y_t}{\partial K_t} = r_t \quad \text{and} \quad \frac{\partial Y_t}{\partial L_t} = w_t$$

where :

r_t – reward for real capital,

w_t – reward for real labour,

then relationship (4.22) can be written as:

$$\frac{\dot{Y}_t}{Y_t} = \frac{T\dot{F}P_t}{TFP_t} + r_t \cdot \frac{K_t}{Y_t} \cdot \frac{\dot{K}_t}{K_t} + w_t \cdot \frac{L_t}{Y_t} \cdot \frac{\dot{L}_t}{L_t} \,. \tag{4.23}$$

Expressions $r_t \cdot \frac{K_t}{Y_t} = \alpha_t^K$ and $w_t \cdot \frac{L_t}{Y_t} = \alpha_t^L$ denote the respective shares of capital and labour in a product. Taking into account the above relations, equation (4.23) takes the form:

$$\frac{\dot{Y}_t}{Y_t} = \frac{\dot{TFP}_t}{TFP_t} + \alpha_t^K \cdot \frac{\dot{K}_t}{K_t} + \alpha_t^L \cdot \frac{\dot{L}_t}{L_t}. \tag{4.24}$$

If labour and capital are the only factors of production, then:

$$\alpha_t^K + \alpha_t^L = 1, \tag{4.25}$$

hence:

$$\frac{\dot{TFP}_t}{TFP_t} = \frac{\dot{Y}_t}{Y_t} - \alpha_t^K \cdot \frac{\dot{K}_t}{K_t} - (1 - \alpha_t^K) \cdot \frac{\dot{L}_t}{L_t}. \tag{4.26}$$

Solow was the first to expand theoretically the above concept of measuring the rate of technological progress (understood as the rate of TFP growth). This equation is also known in the literature of the subject as the Solow residual. From the equation it follows that the rate of technological progress is a difference between the rate of output growth and the weighted sum of the rates of growth of primary factors whose weights correspond to the share that each of the factors contributes to output.

Generalizing the above reasoning to an *n*-factor neoclassical production function with Hicks's neutral technological progress, we have:

$$\frac{dTFP_t}{dt} \cdot \frac{1}{TFP_t} = \frac{dY_t}{dt} \cdot \frac{1}{Y_t} - \sum_{i=1}^{n} \left(\frac{\partial Y_t}{\partial X_{it}} \cdot \frac{X_{it}}{Y_t} \right) \cdot \frac{dX_{it}}{dt} \cdot \frac{1}{X_{it}}, \tag{4.27}$$

Expression $\varepsilon_i = \frac{\partial Y_t}{\partial X_{it}} \cdot \frac{X_{it}}{Y_t}$ denotes the elasticity of the production function with respect to *i*-th factor of production. In other words, the rate of TFP growth is a difference between the rate of output growth and the weighted sum of growth rates of factor whose weights indicate the output elasticity with respect to a given factor.

The rate of TFP growth can actually be estimated by finding output elasticities for each factor. This gives rise to another problem, concerning the form of the production function. On both theoretical and empirical grounds, it is usually assumed that the production function is a neo-classical two-factor function of the Cobb-Douglas type:

$$Y_t = A_t K_t^{\alpha} L_t^{1-\alpha} \tag{4.28}$$

where:

Y_t - the value of output (in constant prices);

A_t - total factor productivity $A_t = TFP_t$

K_t - the value of capital (in constant prices)

L_t - labour inputs.

α, 1-α- the elasticity of production with respect to capital and labour inputs, respectively.

The function (4.28) as written for the rates of growth is:

$$\frac{\dot{Y}_t}{Y_t} = \frac{\dot{A}_t}{A_t} - \alpha \cdot \frac{\dot{K}_t}{K_t} - (1-\alpha) \cdot \frac{\dot{L}_t}{L_t}, \tag{4.29a}$$

hence

$$\frac{\dot{A}_t}{A_t} = \frac{\dot{Y}_t}{Y_t} - \alpha \cdot \frac{\dot{K}_t}{K_t} - (1-\alpha) \cdot \frac{\dot{L}_t}{L_t}, \tag{4.29b}$$

which is equivalent to (4.26).

In other words, the rate of TFP growth is sought by estimating the parameters of the production function (4.28). This can be done by either making a direct estimation of its parameters (introducing additionally variable A_t in the function, usually treated as a function of time), or calibrating its parameters assuming that the long-run elasticities are equal to each factor's share in output (in most cases, for capital the parameter equals the share of gross operating surplus in value added, and 1-α equals the share of labour costs in value added).

4.5.3. Decomposition of total factor productivity

The question arises whether the rate of TFP growth determined as described can be attributed only to the effects of technological progress. Are there any other aspects that can affect the value of TFP growth? Quite plainly, the value of the rate of TFP growth mostly depends on (Nadiri, [1970], pp. 1140-1141):

a) the form of the production function indicating relations between the volume of output and the volume of production factors,

b) the manner of measuring output and production factors, and on taking into account changes in the quality of these factors,

c) addressing factors of production other than labour and capital.

The first postulate actually means that the form of the production function directly affects the level of output elasticities with respect to its factors, *ipso facto* affecting immediately the dynamics of TFP (cf. relation (4.27)).

The way production and its factors are approached is very important for the measurement of both the level and dynamics of TFP. Treating TFP as the level of Hicks's neutral technological progress entails the assumption of the homogeneity of factors and their unlimited substitution. However, both these assumptions are quite unrealistic, and so they have attracted criticism. The assumption of factor homogeneity is omitted because of the concept of embodied technological progress[6]. For capital goods, it is assumed that the technology embodied in new goods is more modern technology and thereby such goods are more productive than older ones. This fact is allowed for by capital vintaging that implies that all available capital is treated as a weighted sum of capital goods of the respective vintages, whose weights indicate the effectiveness of particular vintages (Allen, [1975], pp. 252-254). Then homogeneity only refers to the given generation of capital goods. In addition, the vintage models narrow the range of substitution between factors of production (more on this subject can be found in the works by Allen ([1975], pp. 278-299, 308-313), Fiedor ([1986], pp. 81-95), Juszczak-Schumacher ([1996], pp. 72-90), Tokarski ([2001b], pp. 13-20).

The embodied technological progress may also concern labour. Then it is connected with changes in the qualitative characteristics of some part of the total stock of labour force and, because it leads to productivity of labour, also with the term of human capital (more in the next chapters). The different productivity of particular components of labour in the production function is usually allowed for by constructing one variable that describes both the quantity and quality of labour, or by adding another factor in the production function, representing the quality of labour (Juszczak-Szumacher, [2000], p.148).

The homogeneity of factors can be addressed in the production function by showing the factors as units of effectiveness. Consequently, total factor productivity can be decomposed into effects related to the generally available technical knowledge (A_t^*) and effects arising from better quality of capital and labour applied in the production process (respectively A_t^K and A_t^L). The first of the factors is connected with innovative processes taking place in the economy, which are related to neither specific production apparatus nor labour force resources utilized in the process of production (see Welfe, [2001], p.33 and chapter 1 of this book). Adding these quality factors to a two-factor production function of the Cobb-Douglas type give the function the following form:

[6] The embodied technological progress applies only to parts of the available production factors, it usually contains machinery and equipment installed successively and operation thereof.

$$Y_t = A_t^w (A_t^K K_t)^{\alpha} (A_t^L L_t)^{1-\alpha} \tag{4.30}$$

or as units of effectiveness:

$$Y_t = A_t \overline{K_t}^{\alpha} \overline{L_t}^{1-\alpha}, \tag{4.31}$$

where:

$\overline{K_t} = A_t^K K_t$ – fixed assets expressed as categories of effectiveness (reflecting the growing productivity of capital);

$\overline{L_t} = A_t^N L_t$ – labour expressed as categories of effectiveness (reflecting the improving quality of employees as a result of their education and occupational experience).

Assuming that the above effects of technological progress are separable, production function (4.30) can be written as:

$$Y_t = \underbrace{A_t^w (A_t^K)^{\alpha} (A_t^L)^{1-\alpha}}_{A_t} K_t^{\alpha} L_t^{1-\alpha} \tag{4.32}$$

Naturally, the presented decomposition of total factor productivity is not the only one possible. Decomposition is mainly determined by the way technological progress is treated in the production function and by its factors of production. Therefore, if we add another factor to the production function, e.g. a stock of human capital, then TFP will be adjusted for changes in the quality of employees involved in the production process. The theory underlying this concept was developed by Mankiw, Romer and Weil [1992] who expanded the neo-classical, two-factor production function of the Cobb-Douglas type by adding another factor of production describing the resource of human capital H_t, while assuming that the progress of science and general technical knowledge affects production exclusively through its application of the knowledge by the employees (Manikw, Romer, Weil, [1992], pp. 416-418). Therefore, the function of production in the Mankiw-Romer-Weil model takes the following form:

$$Y_t = K_t^{\alpha} H_t^{\beta} (A_t^w L_t)^{1-\alpha-\beta}. \tag{4.33}$$

Function (4.33) can be transformed into:

$$Y_t = K_t^{\alpha} \left(\frac{H_t}{A_t^w L_t} \right)^{\beta} \left(A_t^w L_t \right)^{1-\alpha}. \tag{4.34}$$

Expression $\frac{H_t}{A_t^w L_t} = \tilde{A}_t^L$ denotes the relative magnitude of human capital, i.e. the amount of capital per unit of effective labour. Then TFP is incompletely denoted as A^*, because it is connected with the capital of knowledge applied by the employees, which can be written without accounting for the effects of capital effectiveness:

$$Y_t = \underbrace{\left(\tilde{A}_t^L\right)^{\beta} \left(A_t^w\right)^{1-\alpha}}_{A_t} K_t^{\alpha} L_t^{1-\alpha}. \tag{4.35}$$

Likewise, the effects of TFP changes in the Lucas model (Lucas, [1988], pp.7-8) are related to changes in the average stock of human capital and to the proportion of time allocated to work needed to produce goods and to provide services.

A slightly different approach was presented by Romer [1990] who attributed TFP growth exclusively to the rising quality of capital used in production. He took account of the effects of improving quality of employees by adding a variable describing the stock of human capital (H_t). The function he suggested has the following form:

$$Y_t = \left(A_t^w K_t\right)^{\alpha} H_t^{\beta} L_t^{1-\alpha-\beta}, \tag{4.36a}$$

that is

$$Y_t = \underbrace{\left(A_t^w\right)^{\alpha}}_{A_t} K_t^{\alpha} H_t^{\beta} L_t^{1-\alpha-\beta}. \tag{4.36b}$$

The above discussion closes with conclusion that the level and dynamics of TFP as well as its interpretation are mainly determined by the extent to which growing factor effectiveness was taken into account, and by the factors specified in the production function that advance science and technical knowledge.

4.5.4. Measures of production in empirical analyses of total factor productivity

Using the production function to find the level and dynamics of TFP is not very complicated, at least theoretically. Still, empirical analyses should pay attention to many important questions. The first of them concerns the form of the production function. The empirical research employs mostly the Cobb-Douglas function, mainly its parameters are easy to estimate. Another crucial question in dealing with empirical analyses of TFP is the choice of appropriate measures of output and its factors.

The two main concepts of TFP measurement can be distinguished. One uses value added as the measure of output and its only factors of production are labour (L) and capital (K) – this is a so-called KL-VA model (Baily, [1981]; Bruno, [1984]). Labour force (L) is usually expressed as the number of employees (their average annual number is usually applied), or sometimes as the time worked (e.g. research on TFP conducted for the USA by the Bureau of Labour Statistics).

As a measure of capital the average amount of fixed assets is usually employed (determined by their amounts as of year beginning and end), while assuming that the dynamics of the stream of fixed assets (i.e. the service of capital) is proportional to the dynamics of their stock, and to the degree to which capital utilization is constant in time (Welfe, [2003], p.105). The assumption about a constant level of capital utilization seems to strong though, because in real world the demand or supply shocks are possible in some periods (Welfe, [2003], p.105). Therefore, together with expanding resources of capital the stream of services it provides decreases, because GDP and the number of employees are declining (Felbur, [1997], p. 51)[7]. This phenomenon makes total factor productivity fall, which process is completely unrelated to technological changes.

The second approach takes advantage of gross output and extends the list of inputs to include intermediate consumption (a KLM-OUT model) (Griliches, Mairesse, [1983]; Jablonski, [1995]), sometimes decomposing them to consumption of energy (E) and of other intermediate goods (M) (a KLEM-OUT model) (Jorgenson, Gollop, Fraumeni, [1987]), or to consumption of energy (E), services (S) and all other intermediate goods (M) (a KLEMS-OUT model) (Gullickson, [1995]; Kern, Wilder, [1997]).

Whether the above models will be applied to estimate TFP mainly depends on the degree of disaggregation of the economic system analysed. If an analysis treats the economy as a whole or by large economic sector (such as agriculture, industry and the tertiary sector), then the KL-VA model with value added (GDP) prevails in empirical research as a measure of output. This is due to the fact that a huge portion of the products consumed by a large sector in form of intermediate goods is produced by the sector itself. Application of the concept of gross output is therefore closely related to the problem of so-called double accounting of production, and so the gross output does not show the actual volume of output in the sector.

As for particular branches of economy (manufacturing some selected, narrow groups of products) the situation is different. In this case, it can be misleading to try to assess the effectiveness of production exclusively in relation to labour input and capital used by this branch to produce. The problem is that quite a large number of products that the branch utilizes as intermediate goods are manufactured outside, so the double accounting question is not so important here as in large sectors, or in economy as a whole. The quality of products used by a given branch as intermediate goods may naturally improve the effectiveness of

---

7 Regarding the economy of Poland, W. Welfe proposed in his research to correct the dynamics of the stream of fixed assets. His proposal increased the growth rate of a stream of fixed assets by the rate of growth of fixed assets utilization.

production as well (Norsworthy, Jang, [1992]; Jablonski, [1995]). This approach gave rise to the development of models such as KLM, KLEM, KLEMS-OUT, first based on the gross output, and then using the notion of sectoral output[8] to circumvent the double accounting of production within a sector.

4.5.5. Effective and potential total factor productivity

The empirical analyses of TFP and TFP interpretations referring to the effects of broadly understood technological progress need to pay attention to the impacts of differences appearing between the effective and potential output. The potential output is understood as a maximal output that can be attained using available factors of production. The earlier discussions about TFP measurement using both index methods and the production function were underpinned by an assumption, not quite realistic, that the observed output equals the potential output. In fact, this assumption seems to be too strong, as output fluctuations can follow from changes in the business cycle bringing about instability of demand, and not from changes arising from the evolution of technology.

A difference between potential output X^{Pot} (output resulting from the production function, assuming full utilization of production capacity) and observed output X_t, expressed as a fraction of the potential output, shows the degree to which production potential $WK_t = \frac{X_t}{X^{Pot}}$ is utilized (symbols after Welfe, [2003], p. 107). Therefore, the observed output is equal to the potential output, adjusted for the degree of utilization of production potential, i.e.:

$$X_t = WK_t \cdot X_t^{POT}. \tag{4.37}$$

Assuming, as empirical researchers usually do, that the production function becomes a Cobb-Douglas function, relation (4.37) can be re-written as:

$$X_t = WK_t A_t K_t^{\alpha} L_t^{1-\alpha}. \tag{4.38}$$

The rate of TFP growth determined by the relation equals:

$$\frac{\dot{A}_t}{A_t} = \frac{\dot{Y}_t}{Y_t} - \left(\frac{\dot{WK}_t}{WK_t} + \alpha \cdot \frac{\dot{K}_t}{K_t} + (1-\alpha) \cdot \frac{\dot{L}_t}{L_t} \right), \tag{4.39}$$

and it is smaller than the rate of TFP growth found using (4.29b) by the impacts of changes in the degree of utilization of the production capacity. If the above

[8] The sectoral output is a magnitude of a gross output of a given sector minus the intermediate use of products manufactured by this sector and passed to this sector. [Glaser, 1993; Jablonski, 1995; Zheng, 2002].

relation additionally allowed for adjustments in the stream of fixed assets concerning the degree of their utilization, then TFP growth determined in this way would only describe the effects of broadly understood technological progress (Welfe, [2003], p. 108). An extensive review of techniques serving the calculation of the rate of TFP growth by means of the potential output and empirical findings for the Polish economy can be found in works by Welfe, ([2001], pp. 116-136 and [2003], pp. 108-110).

The question that is still to be resolved concerns terminology. Total factor productivity is usually established using the observed output, and not the potential one. Given the above discussion, the rate of TFP growth determined as presented expresses not only the effects of broadly understood technological progress. Hence the proposals (Welfe, [2001], p. 124-128) that it should be called the effective (actual or observed) TFP to distinguish it from the TFP (found using the potential output) that expresses only the effects of technological progress.

4.6 Endogenisation of technological progress

4.6.1. Endogenous theory of growth as a basis for endogenisation of technical progress

In the first half of the 20th century, the first formal models of economic growth were created. We should mention here models originating from the Keynesian macroeconomic theory (Harrold-Domar models), as well as the neo-classical models (Solow-Swan models). However significantly different in treating the issues of growth, both the currents accept that the major power driving of long-term economic growth is broadly understood technological progress. Notwithstanding, both types of models treat technical progress as exogenous and this practically means that they do not account for the most important determinant of economic growth.

In the late 1950s and early 1960s first attempts to explain the sources of technological progress were made. (Kaldor, [1957]; Arrow, [1962]; Uzawa, [1965]; Shell, [1966]; Phelps, [1966]). Technological progress was associated with investments involving real capital (Kaldor model of 1957 and its modification in 1962, Arrow model of 1962), stocks of scientific and technical knowledge available in an economy (Shell model of 1966), or human capital (Uzawa model - 1965). However, only at the turn of the 1980s that a fuller explanation of technological progress became available, owing to the development of endogenous growth theory. This theory explains that technological progress comes from planned investment decisions taken by rational economic agents (Tokarski, [1998], p. 271). Consequently, in the technological progress-generating process much weight is attached to particular economies' propensities to invest in R+D sectors and/or education. This is due to the fact that endogenous growth theory assumes technological progress to mainly arise from innovations created by local and international R&D (Romer, [1990], pp. 71–102; Aghion-Howitt, [1998], pp. 22–24) and from accumulation of human capital (Lucas, [1988], p.3-42).

4.6.2. Empirical research using a production function with R&D expenditure

First empirical investigations trying to find sources of technological progress were initiated in 1950s and the major names in this field are Griliches ([1957], [1979]), Schmookler [1966] or Mansfield [1968]. They set out to confirm empirically the hypothesis that technological changes taking place in a given economy result from planned decisions taken by economic agents. Technological changes making factor productivity vary were associated with the economy's R&D potential measured, above all by the amount of cumulated expenditure on R&D activities. These analyses were based on the production function (usually being the Cobb-Douglas function), in which cumulated amounts of R&D expenditure were used as an additional factor of production. In most cases, the analyses concentrated on enterprises or branches, and the results confirmed researchers' hypothesis that a positive relationship existed between the amount of expenditure allocated to innovative activities and the dynamics and/or the level of economic development.

Research intensity clearly grew in the 1980s. Researchers worked on explaining of the clear slowdown observed in the economies of the developed countries in that decade. Blame was placed on lower allocations to research and development (Griliches, [1980]; Griliches and Lichtenberg [1984]; Scherer [1982]; Lichtenberg [1986]; Englander [1988]). The methodology applied then used the production function (of the Cobb-Douglas type) including, in addition to the primary factors, also expenditure on innovation (represented by cumulated R&D expenditure). For the general case, this function can be written as (Clark, Griliches, [1982], p. 14; Griliches [1995], p. 55):

$$Y_t = A_t \cdot F\left(BRK_t, K_t, L_t\right), \quad (4.40)$$

where:

Y_t – the volume of output (constant prices),

A_t – factor representing the disembodied, exogenous technological progress,

BRK_t – cumulated R&D expenditure (constant prices),

K_t – capital stock (constant prices),

L_t – labour.

For a production function specified as presented the cumulated R&D expenditure must be estimated. The reality is that we can usually use information about current outlays on R&D activities (*BR*). In most cases, we assume that their volume is equal to the sum of previous R&D outlays while allowing for depreciation rates, which can be written as:

$$BRK_t = BRK_{t-1} + BR_t - \delta BRK_{t-1}, \tag{4.41}$$

where:

BRK_{t-1} – cumulated R&D expenditure in the previous period (constant prices),

BR_t – R&D expenditure in the current period (constant prices),

δ – the rate of depreciation.

The depreciation rate is usually taken in the range 0.05-0.2 (Griliches, [1980], p. 344; Coe and Helpman [1995], p. 883; Welfe, [2001], p. 141). To find the cumulated R&D expenditure based on (4.41), the level of R&D in the first period t_0 must also be determined too. The most frequent assumption is that this level equals (Griliches, [1980], p. 344; Coe and Helpman [1995], p. 883):

$$BRK_0 = \frac{BR_0}{\sigma + \delta} \tag{4.42}$$

where σ stands for the average rate growth of R&D expenditure throughout the analysed period.

Differentiating relation (4.40) with respect to time, and then dividing both its sides by Y_t, we obtain:

$$\begin{aligned}\frac{dY_t}{dt}\cdot\frac{1}{Y_t} = {} & \frac{dA_t}{dt}\cdot\frac{1}{A_t} + \frac{\partial F}{\partial BR_t}\cdot\frac{1}{F(BRK_t, K_t, L_t)}\cdot\frac{dBRK_t}{dt} + \\ & + \frac{\partial F}{\partial K_t}\cdot\frac{1}{F(BRK_t, K_t, L_t)}\cdot\frac{dK_t}{dt} + \frac{\partial F}{\partial L_t}\cdot\frac{1}{F(BRK_t, K_t, L_t)}\cdot\frac{dL_t}{dt}\end{aligned} \tag{4.43a}$$

or :

$$\begin{aligned}\frac{dY_t}{dt}\cdot\frac{1}{Y_t} = {} & \frac{dA_t}{dt}\cdot\frac{1}{A_t} + \frac{\partial F}{\partial BRK_t}\cdot\frac{BRK_t}{F(BRK_t, K_t, L_t)}\cdot\frac{dBRK_t}{dt}\cdot\frac{1}{BRK_t} + \\ & + \frac{\partial F}{\partial K_t}\cdot\frac{K_t}{F(BRK_t, K_t, L_t)}\cdot\frac{dK_t}{dt}\cdot\frac{1}{K_t} + \frac{\partial F}{\partial L_t}\cdot\frac{L_t}{F(BRK_t, K_t, L_t)}\cdot\frac{dL_t}{dt}\cdot\frac{1}{L_t}\end{aligned} \tag{4.43b}$$

Taking symbols:

$\lambda = \frac{dA_t}{dt}\frac{1}{A_t}$ – the rate of exogenous, disembodied technological progress,

$$\frac{\partial F}{\partial BRK_t} \cdot \frac{BRK_t}{F(BRK_t, K_t, L_t)} = \gamma$$ – the elasticity of production with respect to cumulated R&D expenditure,

$$\frac{\partial F}{\partial K_t} \cdot \frac{K_t}{F(BRK_t, K_t, L_t)} = \alpha$$ – the elasticity of production with respect to capital,

$$\frac{\partial F}{\partial L_t} \cdot \frac{L_t}{F(BRK_t, K_t, L_t)} = \beta$$ – the elasticity of production with respect to labour,

we arrive at:

$$\frac{dY_t}{dt} \cdot \frac{1}{Y_t} = \lambda + \gamma \frac{dBRK_t}{dt} \frac{1}{BRK_t} + \alpha \frac{dK_t}{dt} \frac{1}{K_t} + \beta \frac{dL_t}{dt} \frac{1}{L_t}, \tag{4.44}$$

which after transformation gives:

$$\frac{dTFP_t}{dt} \frac{1}{TFP_t} = \lambda + \gamma \frac{dBRK_t}{dt} \frac{1}{BRK_t}, \tag{4.45}$$

where: $\frac{dTFP_t}{dt} \cdot \frac{1}{TFP_t} = \frac{dY_t}{dt} \frac{1}{Y_t} - \alpha \frac{dK_t}{dt} \frac{1}{K_t} - \beta \frac{dL_t}{dt} \frac{1}{L_t}$.

The above means that the rate of TFP growth depends on the rate of growth of cumulated expenditure allocated to R&D.

An alternative to relation (4.45) can be[9]:

$$\frac{dTFP_t}{dt} \frac{1}{TFP_t} = \lambda + \mu \frac{dBRK_t}{dt} \frac{1}{Y_t} \tag{4.46}$$

Besides, Terleckyj [1974] has shown that if the depreciation rate of cumulated R&D expenditure is close to zero, then the growth of such expenditure can be approximated using the current R&D expenditure spent in t time, that is by BR_t. Consequently, relation (4.46) can be written as:

$$\frac{dTFP_t}{dt} \frac{1}{TFP_t} = \lambda + \mu \frac{BR_t}{Y_t}. \tag{4.57}$$

In other words, the rate of TFP growth is a function of intensity of R&D expenditure defined by the share of current R&D expenditure in output. Parameter

---

[9] This results directly from equation (4.43a), assuming that $\mu = \frac{\partial Y_t}{\partial BRK_t}$. Hence parameter μ denotes the marginal product of cumulated R&D expenditure.

μ corresponds therefore to the rate of return on investments made in R&D (Griliches, [1980], p. 344; Sterlacchini, [1989], p. 1550; Nadiri, [1993], p. 9).

4.6.3. Total factor productivity and R&D expenditure at the level of enterprises and branches

The first estimates of equations (4.46) and (4.47) were calculated in the 1980s. They were applied to research analysing relations between the rate of TFP growth and R&D expenditure, both at the enterprise level (Griliches [1980], [1986]; Griliches and Mairesse [1983]) and in branches of industry (Griliches, [1981]; Griliches and Lichtenberg, [1984]; Scherer [1982]). The elasticity of production towards the rate of growth of cumulated R&D expenditure obtained by the authors was 0.06-0.1. When equation (4.47) was applied, the rate of return on R&D investments ranged from 0.2 to 0.5, but in most investigations it was closer to 0.2[10].

In some of the aforementioned empirical investigations attention paid attention to the fact that productivity growth in a given enterprise (branch) depends not only on the company's (branch's) innovative activity, but also on the activity of the related enterprises (branches). The diffusion of innovation effects among particular enterprises or branches of industry is defined as transfers of R&D outlays (R&D spillovers). Technology can be transferred either directly – by purchasing the intermediate or investment goods, licenses etc. or indirectly, when similar technological solutions based on generally available knowledge are applied. In this context, two main types of technology transfers can be distinguished (Griliches, [1979]):

1. Transfer of technology embodied in the intermediate or investment goods, patents (rent spillovers). The external benefits offered by this transfer arise from the fact that the prices of such goods do not fully reflect the quality of goods resulting from innovations in enterprises (branches) that produce them. This price inadequacy is commonly believed to result from imperfect monopolistic pricing caused by competitive pressures on innovating organizations (Cincera, van Pottelsberghe de la Potterie, [2001], p. 2).

2. Transfer of generally available knowledge and technology (knowledge spillovers) outside formal transactions between businesses, stemming exclusively from observation, learning and copying of knowledge already existing in the economy. The vehicle for this type of transfers is formal and informal contacts between people representing science and production, and between the suppliers

[10] Total factor productivity was estimated using value added.

and the recipients; its effects is exchange of information about new technological and organisational solutions, etc. (Meister, Verspagen, [2004], p.20).

Allowing in empirical research for external benefits provided by the transfer of knowledge and technology requires the construction of a suitable yardstick of such benefits. It is usually assumed that their magnitude is proportional to the stock of knowledge (usually measured using cumulated R&D expenditure) held by all possible sources from which it is derived:

$$BRKspill_i = \sum_{j \neq i} w_{ij} BRK_j \tag{4.48}$$

where:

$BRKspill_i$ - the magnitude of benefits provided by an internal transfer of knowledge and technology to the *i*-th object (a company, a branch of economy, or economy as a whole),

BRK_j - the magnitude of knowledge resources of *j*-th object ($i \neq j$),

w_{ij} - the weight denoting the magnitude of knowledge transferred from *j*-th object to *i*-th object, with $w_{ij} \in (0,1)$.

Depending on the mechanism by which technology and knowledge are transferred, two patterns of assigning weights w_{ij} can be distinguished:

1. When a transfer involves knowledge embodied in goods (rent spillovers) formally exchanged between the suppliers and the recipients, the weights can be defined proportionally to the transaction values. The transactions may involve intermediate goods, investment goods and patents. The weights' values are defined based on information about the volumes of flows of investment goods (Sveikausas, [1981]) or about the flows of raw materials, which in the case of intersectoral analyses can be reflected by the values of input-output coefficients (Terleckyj, [1974], Wolff, Nadiri, [1993], Wolff, [1997]). Regarding the international transfer of technology, a key role is played by the amount of import between particular countries (including import of investment goods), the size of foreign direct investments and flows of patents between different economies.

2. Regarding the transfer of generally available knowledge (knowledge spillovers) its magnitude depends on the technological proximity between the suppliers and the recipients. This technological proximity may result, for instance, from their involvement in innovative activities that have similar structure of R&D expenditures, or patent-related activities (Goto, Suzuki, [1989]; Jaffe, [1986], Verspagen, [1997]).

Table (4.1) provides an overview of the results of selected empirical investigation examining R&D impacts on TFP changes, allowing for the sources and effects of external transfers of technology. All investigations concentrated on companies or branches and used the cross section-time series samples. The models are based on relation (4.46) and they take into account a variable showing the amount of benefits being derived from an external transfer of knowledge and technology. The amount of the benefits is determined using (4.48), and the weight-assigning method depends on the mechanisms of transfer.

The result of investigations focused on enterprises or branches of industry seem to confirm explicitly the existence of a positive relationship between the magnitude of R&D expenditure and the rate of TFP growth. Moreover, benefits arising from the external transfer of technology clearly surpass these provided by organization's own expenditure on innovative activity. In the 1990s similar investigations allowing for foreign transfers of technology were conducted (Mohnen, [1992]; Antonelli, [1994], Bernstein, Mohnen, [1998], Branstetter [2001]). Their conclusions are not basically different from those formulated earlier.

Table 4.1. An overview of selected empirical investigations in the impacts of innovation on TFP changes among enterprises and branches

No.	Author/Authors	Level of aggregation (time period)	Sources of external transfer of technology	Results
1.	Terleckyj (1974)	Manufacturing branches (1954–1969)	Flows of intermediate and investment goods between particular branches	Positive and statistically significant parameters showing the rate of return on R&D outlays, both spent in a given branch and in branches supplying intermediate and goods. The rate of return on external R&D outlays higher than for outlays spent by a given branch.
2.	Sveikauskas (1981)	Manufacturing branches (1959–1969)	Flows of investment foods between manufacturing branches	Positive and statistically significant parameters showing the relationship between R&D outlays and TFP growth. The effects of the spillover of knowledge embodied in capital goods are stronger than effects of R&D undertaken in given branch.
3.	Jaffe (1986)	Manufacturing enterprises (panel data on 432 enterprises, years 1973 and 1979)	Patent flows	Positive and statistically significant relationship between R&D outlays, both financed by a given enterprises and by external enterprises, and the rate of TFP growth. Benefits from the transfer of knowledge are stronger in the high-tech enterprises.
4.	Sterlacchini (1989)	Manufacturing branches (1954–1984)	Flows of intermediate and capital goods and patent flows between branches	Positive and statistically significant relationship between the rate of TFP growth and R&D outlays (incurred in a given branch and transferred from outside), especially those embodied in intermediate and investment goods .

No.	Author/Authors	Level of aggregation (time period)	Sources of external transfer of technology	Results
5.	Wolff & Nadiri (1993)	Manufacturing branches (1947–1977)	Flows of raw materials between branches	Positive and statistically significant relationship between R&D outlays incurred by a given branch and productivity growth in that branch. External transfer of technologies embodied in the intermediate goods significantly contributes to TFP changes, but only for outlays made by the private R&D.
6.	Wolff (1997)	Manufacturing branches (1958–1987)	Flows of intermediate goods between branches	Positive and statistically significant relationship between the rate of TFP growth and R&D outlays (incurred by a given branch and transferred from outside and embodied in intermediate goods). Effects of transfers of external technologies and knowledge are stronger.
7.	Los & Verspagen (2000)	Manufacturing enterprises (1977–1991)	Technological links between particular types of activity formed by patent flows	Positive and statistically significant relationship between R&D outlays incurred by a given enterprise and productivity growth (labour productivity). The external transfer of technology is also important. It is, however, irrelevant, whether the stock of external technology and knowledge is a non-weighted average or estimated using formula (4.52).
8.	Bönte (2004)	Manufacturing branches (1980–1993)	Technological links between branches formed by flows of raw materials	Positive and statistically significant relationship between R&D outlays and the rate of TFP growth. For R&D outlays financed by the private sector its value is

No.	Author/Authors	Level of aggregation (time period)	Sources of external transfer of technology	Results
				considerably higher than for publicly financed R&D. This also applies to effects of technology transfers from other branches.
9.	Świeczewska (2004)	Manufacturing branches (1992–2001)	Flows of raw materials between branches	Positive but statistically insignificant relationship between innovation activity* of particular manufacturing branches and TFP growth. The effects of transfers of technologies from other branches are not significant.

* Innovative activity is described by the share of particular branches' innovation outlays in gross output.

4.6.4. Total factor productivity and domestic and foreign R&D activity – macroeconomic approach

The first empirical investigations conducted on the macro level appeared toward the end of the 1980s. They aspired to explain variations in the level of economic development in particular countries with respect to their technological advancement. One of the first authors was Fagerberg [1987] who, using a cross section-time series sample covering 25 countries in the years 1960-1983 (19 OECD countries, Argentina, Brazil, Hong Kong, South Korea, Mexico and Taiwan), carried out empirical research aimed at verifying the hypothesis that the level of technological development significantly affects economic development. Moreover, the poorer countries (with lower GDP per capita) can reap benefits from outside (via import, diffusion, or imitation) owing to the faster development of richer countries. Therefore, the rate of economic growth in the poorer countries can be exceed that resulting from the rate of growth of their own technological level and cumulation of capital, and, additionally, it can be inversely proportional to the level of economic development in these countries. Fagerberg called this case the catching-up effect (Fagerberg, [1987], pp. 87-88).

Fagerberg assumed that the level of technology can be measured using both the magnitude of outlays allocated to this type of activity (measured by the volume of R&D expenditure) and the magnitude of effects of technological activity in the economy (measured by the number of implemented patents). Indices of technological activity are defined as follows:

$$i_{B+R} = \frac{B+R}{Y} \tag{4.49a}$$

$$i_{PAT} = \frac{PAT/N}{EX/Y} \tag{4.49b}$$

where:

$B+R$ – the amount of R&D expenditure in the civilian sector of a given economy (constant prices),

EX – gross export (constant prices),

N – the number of population,

PAT – the number of patents registered in a given economy and implemented outside,

Y – GDP of a given economy (constant prices).

Because of the lack of appropriate statistical data, the Fagerberg model took into account only the second type of the technological activity indices. It results corroborate the hypothesis that the magnitude of technological potential of the given economy positively affects the rate of economic growth. In addition, for the small and medium-sized economies this effect was slightly stronger than in case of OECD countries or all the countries together. Unfortunately, Fagerberg's pioneering efforts omitted the explicit benefits derived from the transfer of foreign technology. They only confirm that economies more committed to technology development have a higher rate of economic growth, resulting in a stronger convergence effect.

The first empirical research investigating the impact of broadly understood innovative activity of economies on their higher effectiveness and allowing for the effects of absorption of foreign technologies was initiated in the early 1990s (Mohnen, [1992]; Park, [1993], Soete, Verspagen, [1993]). However, its results were not explicit. Although most of them confirm an important, positive effect of both domestic and foreign R&D expenditure on productivity growth, but the power of this relationship is different. In the Mohen's research [1992], the elasticity of production with respect to foreign R&D expenditure (embodied in important goods) is higher than in the case of domestic R&D expenditure, but the rate of return on foreign R&D expenditure is around ten times as small as that on domestic expenditure. Park's results [1993] show a stronger impact of foreign R&D expenditure than of domestic expenditure and the effect is inversely related to the size of the economy (it is stronger in smaller economies). In contrast, the research conducted by Soete and Verspagen [1993] did not reveal impacts of foreign R&D expenditure on productivity increases.

Coe and Helpman research [1995] exploring the influence of cumulated domestic and foreign R&D expenditure on the level of TFP takes advantage of a cross section-time series sample containing 22 countries (21 OECD countries and Israel) in the years 1971-1990. As assumed by the authors, the main channel of foreign technology transfers is trade between economies, and the magnitude of this absorption is proportional to the volume of a given country' import, so:

$$\ln A_{it} = \alpha_0 + \alpha_1 \ln S_{it}^d + \alpha_2 m_{it} \ln S_{it}^f, \tag{4.50}$$

where:

A_{it} – the level of total factor productivity in the i-th country in period t;

S_{it}^d – the magnitude of cumulated domestic expenditure on R&D (constant prices);

S_{it}^f – the magnitude of cumulated foreign expenditure on R&D (constant prices)

m_{it} – i-th country' share of the import of goods and service in GDP,

α_1, α_2 – TFP elasticity towards domestic and foreign stock of R&D, respectively.

The cumulated magnitude of R&D expenditure spent abroad was calculated using the formula:

$$S_{it}^{f} = \sum_{j \neq i} \frac{m_{ij}}{m_{i\bullet}} \cdot S_{j}^{d} \tag{4.51}$$

where:

m_{ij} – the volume of import of goods and services to the *i*-th country from a *j*-th country,

$m_{i\bullet}$ – the total volume of import from the *j*-th country to the *i*-th country, $(j = 1,2,.....,21)$; $m_{i\bullet} = \sum_{j} m_{ij}$

S_{j}^{d} – cumulated R&D expenditure in *j*-th country

Formula (4.51) assumes that the amount of foreign expenditure on R&D in the economy is the higher, the more exports come from the country, in which the cumulated R&D expenditure is large enough. The index of foreign R&D expenditure constructed in this way only shows the directions of knowledge and technology transfer, not its intensity.

The TFP elasticity with respect to foreign R&D expenditure in model (4.50) that takes into account the intensity of import of a given country (import's share in GDP) results from author's assumption that the absorption of knowledge and technology is the higher, the more open of a given economy is. Additionally, the TFP elasticities in model (4.50) were related to domestic R&D outlays in such a way so that the size of economies could be taken into account, while distinguishing the group of the richest countries – G7.

The results Coe and Helpman obtained corroborate that a positive relationship exists between the level of TFP and the magnitude of domestic and foreign R&D outlays. The elasticities they calculated for the domestic R&D expenditure amounted to 0.078-0.097, with the value being much higher for the G7 countries (0.134-0.156). The TFP elasticity towards the foreign R&D expenditure varied between 0.06 and 0.092 (values of parameter α_2), and after import was allowed for (TFP elasticity with respect to variable $m_{it} \ln S_{it}^{f}$) it went up to 0.294. In other words, the level of TFP in small economies is more affected by the transfer of foreign knowledge and technologies, and the benefits from this transfer are the higher, the larger openness of a given economy.

Coe, Helpman and Hoffmaister extended the range of research to the group of developing countries (between 1971 and 1999 they researched 77 underdeveloped or developing countries in Africa, Asia, and Latin America).

They aimed at establishing the extent to which the level of productivity in the developing countries depends on the absorption of knowledge and technologies from the developed nations (including the OECD countries and Israel)[11]; then, whether a relationship can be found between a given economy's openness (as measured by the share of import of machinery and equipment from the developed countries in GDP) and the level of developing countries' productivity; and finally whether the absorption of foreign technology is related to the stock of human capital available in the developing countries. Initially, their model was an extension of the Coe and Helpman model of 1995, with additional variables accounting for the openness of the given economy and available human capital (Coe, Helpman, Hoffmaister, [1997], p. 137):

$$\ln A_{it} = \alpha_i^0 + \alpha_i^s \ln S_{it} + \alpha_i^M M_{it} + \alpha_i^E E_{it} + \alpha_{it}^T T_t + \mu_{it}, \tag{4.52}$$

where:

A_{it} – TFP level in the i-th country in period t,

S_{it} – cumulated foreign expenditure on R&D provided by the developed countries covered by the investigation, with $S_i = \sum_{k=1}^{22} \varphi_{ik} S_k^d$, where φ_{ik} is the share of import of machinery and technology from the developed k-th country (k=1,...,22) to the developing i-th country in the total import of machinery and equipment of the i-th country from the developed countries,

M_{it} –the index of an economy's openness as measured by the relation of total import of machinery and equipment to the i-th country from the analysed developed countries, to the level of GDP in the i-th country,

E_{it} – the index of human capital quality in the given country defined as a share of secondary school students in the general number of youth at secondary school age,

T_t – time trend,

μ_{it} – error term.

The domestic R&D expenditure is omitted in this model because the authors believe that the underdeveloped or developing the expenditure is so low that its influence on the productivity of the given economy is insignificant (Coe, Helpman, Hoffmaister, [1997], p. 137)[12].

---

[11] This is the same sample of countries that was used in the previous research Coe and Helpman [1995].

[12] The domestic R&D expenditure was omitted also because of unavailability of complete empirical data on R&D expenditure in the investigated group of developing countries.

To verify a hypothesis that the degree of economy's openness and the quality of its human capital are conducive to the absorption of foreign knowledge and technology, the authors modified model (4.25), transforming it to be:

$$\ln A_{it} = \alpha_i^0 + \alpha_i^s \ln S_{it} + \alpha_i^M M_{it} + \alpha_i^E E_{it} + \alpha_i^{SM} M_{it} \ln S_{it} + \alpha_i^{SE} E_{it} \ln S_{it} + \alpha_{it}^T T_t + \mu_{it} \quad (4.53)$$

In this model, parameter α_i^{SM} denotes the power of interaction between the degree of economy openness and the importance of benefits derived from the transfer of foreign technologies. In other words, if this parameter is positive, then the relation between foreign R&D and productivity of the given economy is the higher, the larger the degree of the given economy openness to trade relations with the developed countries (as measured by M_{it} index). Likewise, a positive value of parameter α_i^{SE} means that the influence of foreign technology transfers is the larger, the higher the qualifications of employees in a given economy.

The results obtained by the authors corroborate that the level of productivity in the underdeveloped or developing countries is largely determined by the transfer of technologies from the developed countries. A 1% increase in the cumulated R&D expenditure in the developed countries makes the developing countries'TFP grow by almost 0.06% on average. The intensity of trade between two classes of countries and the quality of their labour force are also important for the level of their TFP. The research additionally confirms that the import of machinery and equipment is a very important channel of technology transfer from the developed countries[13].

The research conducted by Coe and Helpman and Coe, Helpman and Hoffmaister allows concluding that the main channel of transfer of foreign knowledge and technology is trade relations between individual countries. The power of these relations is expressed by the values of coefficients describing weights used to determine the magnitude of foreign cumulated R&D expenditure (cf. (4.55)). This rather orthodox view on the possible channels of transfer of foreign technology formulated by these authors was challenged by Keller [1998]. He presented the estimation results of Coe and Helpman model, applying an alternative measurement of the cumulated foreign R&D expenditure. He considered two methods for calculating variable S_{it}^f. In the first case, he calculated the amount of cumulated foreign expenditure on R&D as a weighted average of cumulated R&D expenditure in countries being trade partners of the

---

[13] Coe, Helpman and Hoffmaister tested different channels of foreign transfers of technology. They assumed three possible types of transfer, namely: import of machinery and equipment, import of manufactured products and import of all goods and service. The best effects were obtained, however, for the import machinery and equipment.

given country, assessing the weights at random. Under the other method, he assumed the value of variable S_{it}^{f} was a non-weighted sum of foreign cumulated R&D expenditure. The TFP elasticities with respect to foreign R&D expenditure that Keller obtained were higher than Coe and Helpman (α^{f}=0.329 in the first method of calculating S_{it}^{f} (R^2=0,747) and α^{f}=0.335 in the other case (R^2=0,748)). Keller confirmed that trade relations between individual countries were important for the transfer of technologies embodied in the imported goods, stressing at the same time that the transfer of disembodied foreign knowledge and technology was also significant for the level of development of the given economy[14].

The results obtained by Coe and Helpman gave an impulse to starting more investigations into the role played by the transfer of foreign technologies in the process of economic development in both the developed countries and the developing ones. These investigations concentrated on the methods for measuring the stock of knowledge and technology from abroad, allowing for the different channels of their transfer. One of the first authors to take on such research were Lichtenberg and van Pottelsberghe de la Potterie [1998]. Their explorations built on the model proposed by Coe and Helpman [1995], but they modified the way of measuring the magnitude of foreign R&D expenditure. Their research was induced by the observation that the measure of cumulated R&D expenditure as proposed by Coe and Helpman (4.51) was very sensitive to changes in the level of aggregation. They proposed an alternative measure of the expenditure on cumulated foreign R&D, taking into account both the direction and intensity of such a transfer (Lichtenberg, van P. Potterie, [1998], p. 1485):

$$S_{it}^{f} = \sum_{j \neq i} \frac{m_{ijt}}{y_{jt}} \cdot S_{j}^{d} \qquad (4.54)$$

where y_{jt} is GDP volume of the j-th country in period t. Therefore, the magnitude of a transfer to a given economy is proportional to the volume of imports in this economy and to the intensity of R&D expenditure in countries exporting goods and services to the given economy. Lichtenberg and Pottelsberghe de la Potterie conducted their research using the same cross-section-time series sample as Coe and Helpman [1995]. Although slightly better in statistical terms, their results led to similar conclusions. The effects of domestic expenditure on R&D are stronger

---

14 Responding to Keller research, Coe and Hoffmaister demonstrated that the "randomly" picked weights that Keller used to estimate the volume of foreign cumulated R&D expenditure were actually not random, and that the value of variable S_{it}^{f} estimated by Keller by means of the first method was a regular arithmetic mean with error term. In addition, they proved that for really "randomly picked" weights the transfer of foreign knowledge and technology would not significantly affect the TFP level (Coe, Hoffmaister, [1999], pp. 12-13).

in large economies, but the influence of foreign expenditure is the stronger, the larger openness of a given economy is (determined by trade relations).

One of the first authors to deal with a much wider spectrum of possible channels of technology transfers between individual economies were Xu and Wang ([1999], [2000]), and Lichtenberg and Pottelsberghe de la Potterie ([1996], [2001]). In addition to taking into account the import of goods and services, they also paid attention to the import of investment goods (machinery and transport equipment [Xu, Wang, 1999 and 2000]), foreign direct investments (FDI) broken down into inward FDI and outward FDI (Lichtenberg and van P. Potterie, [1996], [2001], Xu and Wang, [2000]). Additionally, Xu and Wang stressed the importance of the transfer of disembodied knowledge and technologies from abroad for TFP.

The first important investigations into FDI effect of economy's effectiveness viewed in terms of productivity growth come from the 1970s[15]. These investigations mainly explored sectors or enterprises and aimed at explaining labour productivity variations by means of the volume of foreign direct investments (FDI) flowing into the sectors (Caves, [1974], Blomström, [1986]). Their results seem to corroborate the hypothesis of a positive contribution of foreign capital to the growth of productivity.

The next investigations into FDI as a factor increasing productivity (at an enterprise level and macro economic level) did not provided such explicit results. They show that the inward FDI benefits only occur in companies where the foreign capital is involved[16] (Aitken, Harrison, [1993]) or in companies (sectors) that are technologically relatively close to foreign companies (Kokko, [1996], Sjohölm, [1997]). The FDI is viewed as a factor directly affecting the level and/or rate of productivity growth. Should the FDI be treated as one of many possible channels of international transfers of knowledge and technology, then the volume of FDI can be used – as in the case of import – to estimate the cumulated expenditure on foreign R&D[17]. FDI was used to find the level of absorption of foreign knowledge in the work by Lichtenberg and Pottelsberghe de la Potterie [1996]. Using information about the volumes of foreign capital flowing particular

[15] In addition to ensuring flows of financial capital, foreign direct investments are a source of different types of knowledge and physical and human capital; thus they can determine economic growth of the recipient country. This thesis seem to be confirmed by research conducted by Romer [1993], Borensztein, Gorgio and Lee [1998],which showed a positive relationship between the volume of FDI and economic growth in the developing countries.

[16] The results of these investigations seem to suggest that transfers of modern technologies from abroad take place between the parent company and its subsidiary under restricted diffusion among domestic firms.

[17] Assuming that benefits offered by R&D expenditure is proportional to the volume of flows of foreign capital represented by direct investments.

countries[18] – both inward and outward[19] – the authors determined the magnitude of cumulated expenditure on foreign R&D as:

$$S_{it}^{ff} = \sum_{j \neq i} \frac{f_{ijt}}{k_{jt}} \cdot S_j^d \tag{4.55a}$$

and

$$S_{it}^{ft} = \sum_{j \neq i} \frac{t_{ijt}}{k_{jt}} \cdot S_j^d , \tag{4.55b}$$

where:

f_{ijt} –FDI flowing into *i*-th country from *j*-th country in period *t*,

t_{ijt} – FDI placed by *i*-th country in *j*-th country in period *t*,

k_{jt} –gross outlays allocated to fixed assets in *j*-th country in period *t*.

They also addressed the significance of import as another source of transfer, using the relation (4.45) to estimate the amount of cumulated expenditure on R&D. Following Coe and Helpman [1995], they assumed that the level of TFP in a given country (F_{it}) was a function of cumulated domestic R&D expenditure (S_{it}^d; they distinguished the group of the richest countries G7) and foreign cumulated R&D expenditure (S_{it}^f)) and tested a large range of possible channels of transfer. The obtained results show that the major channels of technology transfers are import and only outward direct foreign investments. On the other hand, it was not corroborated that the inward transfer of technology related to the inflow of foreign capital to a given country also provides benefits.

Comparable results were obtained by Xu and Wang [2000] who used a similar model to that formulated by Lichtenberg and Pottelsberghe de la Potterie (they additionally took account of human capital and an index describing the volume of transfer of disembodied stock of knowledge)[20] and assumed that the transfer of knowledge between particular countries could be mainly effected via the import of machinery and equipment and flows of capital represented by direct investments. Their results confirm that the major channel of knowledge transfer is the import of capital goods and direct investments that the given economy had made abroad. The diffusion of disembodied technology also seems to be important for economy development. On the other hand, assumptions that direct

[18] The investigations were based on a group of 13 OECD countries in the years 1971-1999.

[19] The information about FDI volumes invested abroad by the given economy was used because the authors' belived that a direct access to modern foreign technologies was much more effective.

[20] They determined the magnitude of benefits derived from the transfer of disembodied knowledge as a non-weighted sum of cumulated expenditure on R&D in the other countries covered by the research, i.e. $S_i^f(UW) = \sum_{j \neq i} S_j^d$ (Xu, Wang, [2000], p. 6).

foreign investments flowing into an economy are a significant vehicle for foreign technologies were not confirmed.

The results obtained by Hejazi and Safarian [1999] seem to be a little different. Investigating the specification of the international channels of innovation diffusion, they assumed that only direct investments from the most technologically advanced countries could make an economy grow. The foreign R&D expenditure was defined as a weighted average of cumulated outlays on this type of activity in six largest economies. The pattern they used to assign weights followed Coe and Helpman's proposals (4.51), i.e. import was replaced by the volume of FDI provided by the most developed countries of the world. The results confirmed the assumptions. Moreover, the transfer of technology embodied in FDI stimulated TFP growth more strongly than the transfer of technology embodied in imported goods[21].

The above investigations into the role played by foreign direct investments in the process of international diffusion of innovations were concentrated only on the developed countries. In the recent years, also the less developed or developing countries were covered by the same type of analyses. In this context Crispolti and Marconi's [2005] investigations are worth mentioning. Using a sample 45 less developed or developing countries – mainly from Africa, Latin America and Asia – they set out to answer the question if, and if yes, to what extent, the productivity growth in these countries was related to R&D expenditure made in the most developed countries of the world (the USA, Japan and countries in Western Europe). They assumed that technologies were transferred via import and inward FDI. They omitted the domestic expenditure on R&D assuming, as Coe, Helpman and Hoffmaister [1997] did, that in case of less developed countries this sphere was of marginal importance for creation of new technologies. However, they took account of the factor indicating the stock of human capital, which can considerably determine the absorption of foreign technologies. The results they obtained confirm that in the less technologically developed economies both the mentioned channels of technology transfers are significant for productivity growth, with the benefits derived from these transfers being the bigger, the larger stock of human capital.

It is worth stressing that a vast majority of empirical investigations exploring the R&D effect on productivity growth in the macroeconomic context use the cross section-time series sample. Only some of them target individual economies, even of the most developed countries. Investigations limited to one economy are usually conducted at the level of enterprises or a branch (Hanel,

[21] The elasticity of TFP with respect to foreign R&D expenditure embodied in FDI amounted to 0.136 and, with respect to foreign R&D expenditure embodied in imported products, 0.106.

[2000]; Cameron et al., [2005]). An exception is the Jorgenson research[22] (Jorgenson, Dougherty, [1996]; Jorgenson [2001]) on the US economy, where he tried to relate economic growth to information technologies (IT). In this context, the research conducted on the Polish economy by W. Welfe's team is worth mentioning. Based on a time sample spanning the years 1970-1998 they tried to relate total factor productivity (more precisely, the productivity of machinery and equipment) to cumulated R&D expenditure (both domestic and in OECD countries being Poland's main trading partners[23]), with the magnitude of absorption of foreign technologies being proportional to the share of investment import in the domestic expenditure on investment in machinery and equipment. The model they used in their research is as follows:

$$\ln A_t^K = \lambda_1 \ln BIRKS_t + \lambda_2 \left({M7_t}/{JV_t} \right) \ln BIRMS_t , \tag{4.56}$$

where:

A_t^K – total productivity of machinery and equipment in period t;

$BIRKS_t$ – cumulated domestic expenditure on R&D in period t,

$BIRMS$ – cumulated foreign expenditure on R&D spent in period t,

$M7_t$ – the investment import (constant prices),

JV_t – investment expenditure on machinery and equipment (constant prices).

The relation (4.56) was used in the Cobb-Douglas function of production[24]. The obtained estimates of the parameters λ_1 and λ_2 (λ_1=0.15 and λ_2=0.17,respictevely, in the variant assuming that the measure of capital is the stock of machinery and equipment, and λ_1=0.43 and λ_2=0.092 in the variant with capital measured using all available fixed assets). In the first variant, the elasticity of TFP with respect to the country's cumulated expenditure on R&D is similar to the elasticity obtained by Coe and Helpman [1995] for the OECD countries, especially regarding the industrial countries, and in the second variant the elasticity is comparable with that obtained for the G7 countries. In the case of foreign R&D expenditures, the elasticities are considerably below those obtained

---

[22] More in the next section of this chapter.

[23] The research took into account six OECD countries: France, the Netherlands, Germany, USA, UK and Italy.

[24] This function has the following form (W. Welfe, [2001], pp. 146-147):
$X_t = B(K_t A_t^K)^{\alpha} (N_t A_t^N)^{1-\alpha} WKZ_t^{\mu} WN_t^{u} e^{\varsigma_t}$, where X_t – the volume of GDP, K_t – the stock of capital as measured by the stock of machinery and equipment (variant 1) or by the stock of all gross fixed assets (variant2), N_t – the number of employees, A^K – technical quality of equipment, A^N – the quality of the active employees, WKZ – the change utilization index, WN – the working time index, ξ – an error term.

by Coe and Helpman [1995] for the developed countries. According to the results, the importance of the domestic R&D sphere for the productivity growth in the Polish economy is comparable with the weight in the developed countries. Regarding the absorption of foreign technology, such a low level of elasticity may suggest that the import of machinery and equipment did not include the state-of-the-art devices (Welfe, [2001], pp. 147-148).

As far as macroeconomic research in the total factor productivity in the Polish economy is concerned, the results of L. Zienkowski investigations [2003] are worth mentioning[25]. In his research, Zienkowski tried to explain variations in the level of TFP using not only the determinants of technological progress (i.e. R&D outlays or human capital) but also factors related to system changes (at least to the restructuring and privatisation of the public sector). Even through the results having a rather contributory character as the authors themselves stress, they indicate the importance all specified factors for explaining the variability of TFP.

4.6.5. Total factor productivity and the effects of R&D

The sphere of R&D "produces" innovation. One of such "products" is information technologies (ICT) encompassing computer and communication devices equipment and software. The unquestionable leader in production of information technology is the USA. Numerous investigations confirm the importance of ICT contribution to country's economic growth, among which those conducted by Jorgenson [2001] should be mentioned. His investigation mainly concentrated on the specification of the US sources of economic growth in the post-war period, with special attention paid to ICT. The reserch methodology he applied takes advantage of the concept of a production possibility frontier (Jorgenson et al., [1996]) describing an effective combination of inputs and outputs for an entire economy. The macroeconomic production function Y containing the production of investment and consumer goods (disaggregated into ICT products and others) is a function of capital stock (disaggregated into capital representing information technologies and other capital), labour and TFP (Jorgenson, [1996], pp. 27–28):

$$Y(I_n, I_c, I_s, I_t, C_n, C_c) = AX(K_n, K_c, K_s, K_t, L), \tag{4.57}$$

where:

[25] The authors considered three groups of factors in their research – reforms (represented by the private sector' share in value added), modernisation (the cumulated expenditure on future development (i.e. outlays on fixed assets, education and R&D)), foreign direct investments, and the share of employees with secondary and tertiary education in total employment), as well as a variable representing the development gap (measured by the Polish GDP per capita to the EU average).

I_n – investments in goods not containing information technologies,
I_c – investments in computer equipment,
I_s – investments in software,

I_t – investments in communications equipment,
C_n – consumption of goods not containing information technologies,
C_c – consumption of goods containing information technologies,

K_n – capital stock unrelated to information technologies,
K_c – capital stock represented by computer equipment,
K_s – capital stock represented by software,
K_t – capital stock represented by communications equipment,
L – labour force,
A – total factor productivity.

Assuming that products and factors of production are perfectly competitive and that production shows constant returns to scale, relation (4.57) can be written in terms of growth:

$$\begin{aligned}&\overline{w_{I_n}}\Delta\ln I_n+\overline{w_{I_C}}\Delta\ln I_c+\overline{w_{I_s}}\Delta\ln I_s+\overline{w_{I_t}}\Delta\ln I_t+\overline{w_{C_c}}\Delta\ln C_{C_c}+w_{C_n}\Delta\ln C_n=\\&=\Delta\ln A+\overline{v_{K_n}}\Delta\ln K_n+\overline{v_{K_c}}\Delta\ln K_c+\overline{v_{K_s}}\Delta\ln K_s+\overline{v_{K_t}}\Delta\ln K_t+\overline{v_L}\Delta\ln L\end{aligned}\tag{4.58}$$

where:

Δ lnx – the rate of growth of a variable x (respectively: $I_n, I_c, I_s, I_t, C_n, C_c, K_n, K_c, K_s, K_t, L$)

$\overline{w}$ – average share of output of a given good in total output

$\overline{v}$ – average share of each factor in total output.

At the same time, $\overline{w_{I_n}}+\overline{w_{I_c}}+\overline{w_{I_s}}+\overline{w_{I_t}}+\overline{w_{C_n}}+\overline{w_{C_c}}=1$ and $\overline{v_{K_n}}+\overline{v_{K_c}}+\overline{v_{K_s}}+\overline{v_{K_t}}+\overline{v_L}=1$. In other words, the rate of output growth being a weighted average of the rates of growth of the manufactured investment and consumer goods is equal to the weighted average of the rates of growth of primary factors and the rate of TFP growth, assuming a full utilisation of factors of production.

TFP can be further decomposed into total productivity of factors containing information technologies (A_{IT}) and total productivity of the remaining factors (A_n), so (Jorgenson, [2001], p. 22):

$$\Delta\ln A=\overline{u_{IT}}\Delta\ln A_{IT}+\overline{u_n}\Delta\ln A_n \tag{4.59}$$

Jorgenson explored the US economy between 1948 and 1999. The results of his research explicitly indicate that the contribution of information technology to economic growth was growing. For instance, an average annual output growth between 1948 and 1973 stood at 3.99%, of which only 0.2 of a percentage point was the growth of ICT products. However, between 1995-1999 production grew at 4.08% a year, in which value 1.81 p.p. was contributed to information technologies. A similar trend can be observed with respect to TFP. In the 1990s the TFP growth was mainly drive by the growth of TFP embodying modern technologies.

Colecchia and Schreyer [2002], who conducted similar research in the group of nine OECD countries between 1980 and 2000[26] arrived at similar results. In the first decade of the period in which the ICT share was analysed, production growth was country-specific and varying from 0.2 to 0.5 p.p. In the 1990s, this numbers went up to 0.3-0.9 p.p., and in both periods ICT was the highest in the USA.

Another research that is worth mentioning is the research conducted by Leeuwen and Weil for Holland [2003]. Using panel data (on almost 8,000 service-providing companies between 1994 and 1998), they analysed the effect of ICT on the growth of labour productivity. They formulated a hypothesis that higher productivity of labour was induced not only by a growing capital-labour ratio (they decomposed capital into capital embodying information and communication technologies and other capital), but also by ICT transfers from outside[27]. Their results show that ICT significantly contribute to the growth of labour productivity and its effect is larger in the case of external technologies[28].

When R&D effects are being discussed, its other "product", i.e. patents, should not be omitted. Indicators derived from patent statistics such as the number of inventions filed and granted patents are among the major measures of innovation. Patent statistics with patent classification by their type and patent application to particular sectors of economy are quite popular in determining the potential channels of the intersectoral transfer of knowledge and technology (a short overview of such investigations is provided table 4.1). Information on the international flows of patents can help identify the potential channels of knowledge flows between particular economies. In addition, the application of

[26] The sample contained the following countries: Australia, Canada, Finland, France, Germany, Italy, Japan, UK and the USA.

[27] This transfer is measured by the sum of capital stocks embodied in information technologies in the other enterprises covered by the research.

[28] The elasticity of labour efficiency with respect to own ICT technologies is 0.029, and 0.079 for the outside technologies (Leeuwen, Weil, [2003], p. 28).

new and constantly improved indicators based on patent statistics (i.e. patent citations) makes it possible to identify the potential channels of knowledge transfers from abroad (i.e. knowledge spillovers). However, the disadvantage of using patent statistics (indicating both patent applications and patent citations) to find the paths of international diffusion is that not all available knowledge is protected by patent law, because patent application or citing depends on the enterprise's strategic decision. The number of patent applications varies from country to country, as the incentives to patent are country specific. Hence, one of the biggest drawbacks of the patent statistics is its low comparability.

Empirical research on patent application as a way of identifying the paths of the international transfer of knowledge is still much less frequent than investigations into the transfer of knowledge and technology based on direct relations (mainly trade relations) between the supplier and the buyer. However, the results of research headed by Jaffe (Jaffe, Trajtenberg, [1999]) that aimed, *inter alia*, at constructing an appropriate measure based on the available patent statistics that could be used to find the paths as well as the power of knowledge spillovers between particular economies. The possible spillover paths are identified using the probability of patent citations, whose empirical measure is the actual frequency of such citations[29]. The authors' conclusions seem to confirm their hypothesis that the geographical proximity is important for the international flow of patents. The empirical frequency of patent citations can be an alternative measure – beside the proximity of trade relations – used to determine the amount of expenditure on foreign R&D[30].

4.7 Summary and conclusions

The theoretical concepts underlying the measurement of technological progress and its endogenisation can be summed up as follows.

Firstly, the macroeconomic effects of technological progress are usually considered in the context of a neoclassical, two-factor production function. They are usually equated with a shift in the function's isoquant towards the origin of the coordinate system, which means that a given output can be obtained at smaller inputs or, which amounts to the same thing, it is possible to obtain a larger output using the same inputs. The effects of technological progress can therefore be expressed by changes in the particular parameters of this function or its chosen characteristics.

[29] The empirical frequency of patent citation is calculated as a ratio of the total number of citations of the *j*-th country's patents in the *i*-th country *i* to the product of patent numbers in the two countries (Jaffe, Trajtenberg, [1999], pp. 108-109).

[30] Jaffe's results were also corroborated by MacGarvie research [2005] that exploring the channels of international knowledge spillovers.

Secondly, at the macroecomic level of technological progress is based on the assessment of effectiveness of the analysed economic structures. Changes in effectiveness are usually expressed via changes in the structure of output itself, in the structure of factors and that relations that link them, as well as changes in factor productivity (or absorbability). Productivity (absorbability) may relate to one factor of production (i.e. labour productivity or capital productivity) or to a combination of factors (i.e. the total factor productivity - TFP).

Thirdly, total factor productivity considered in the context of neoclassical production function describes the amount of output per unit of combined factors of production. TFP changes describes by shift of this function at a given level of factors of production. This mechanism is graphically represented by the isoquant of the production function moving towards the origin of the coordinate system, with the shares of labour and capital remain unchanged. Theoretically, TFP shows the rate of neutral technological progress as defined by Hicks.

Fourthly, the rate of TFP growth analysed in respect of the neoclassical function of production is the difference between the rate of output growth and the weighted average of the rates of growth of factors, with the factors' weights being elasticities of production towards particular factors. In line with Clark's marginal theory, these elasticities are equal to the share each production factor contributes to output. This relation is also known as the Solow residual equation.

Fifthly, there are two alternative ways of seeking total factor productivity. The first uses the index methods and the other one employs the neoclassical function of production. The index methods of seeking TFP are based on the aggregated statistical indices (as formulated by Laspeyres, Paasche and Fischer) or the Divisia's index and its discrete approximation (the Törnquist index). Estimation of TFP consists in finding the parameters of a production function with neutral technological progress as defined by Hicks, and applying the Solow concept of residuals.

Sixthly, technological progress has always been treated as the major determinant of economic growth. However, the first attempts to endogenize it, i.e. to explain the sources of technological progress, were undertaken at the turn of the 1950s. Only the development of endogenous theory of economic growth that took place in the late 1980s made it possible to account fully for the sources of technological progress. The endogenous models treat technological progress as the effect of planned investment decisions of rationally behaving economic agents. Hence, the propensity to investment in R&D and education is given considerable importance in the process of generating technological progress. It is assumed that one of the sources of technological progress is innovation, increasingly modern technical solutions (resulting from R&D) and stock of human capital reflected in the rising quality of employees.

Seventhly, the first empirical investigations trying to endogenise technological progress were undertaken already in the 1950s. Researchers tried to relate technological changes expressed by variations in factor productivity to economies' R&D potential, which was usually measured by the volume of cumulated expenditure on R&D. The investigations mainly focused on enterprises or branches in the economy, and their results confirm the formulated hypothesis of a positive relationship between the amounts allocated to innovation and the pace and/or the level of technical development.

Eighthly, the first empirical investigations in the endogenisation of technological progress paid attention to the fact that the growth of productivity in an enterprise or a branch of economy (mainly industrial) depends on the enterprises's (branch's) technological activity, but also on the activity of the related enterprises (branches). The literature of the subjects defines the diffusion of innovations effects among particular enterprises or branches as R&D spillovers. Technology can transferred between particular enterprises (branches) either direct, via transactions involving intermediate or investment goods, licenses, etc., or indirectly, when enterprises (branches) apply similar technological solutions based on generally available knowledge.

Ninthly, macroeconomic empirical investigations into the endogenisation of technological progress were undertaken in the late 1950s. They related the effects of technological progress (usually interpreted as productivity changes) to the domestic R&D activity (expressed in terms of current or cumulated R&D expenditure), R&D effects (the number of patents and the stock of ICT) and human capital. The transfer of knowledge and technology from abroad was also important. This transfer may use direct relations between economies (flows of intermediate goods, capital goods, import of goods and services, FDI) and indirect relations (flows of patents). The outcomes of empirical investigations into the possible channels of transfer of foreign knowledge and technology do not indicate clearly, which of the mentioned transfer methods is the most effective. Besides, in many cases the appropriate way of diffusion depends on availability of statistical data that could be used in the research.

Tenthly, the above theoretical deliberations suggest that the macroeconomic model of Poland's knowledge-based economy should consider a possibly full range of factors representing the embodied capital of knowledge. Among these factors, the capital related to domestic and foreign R&D activity are worth mentioning. The critical aspects is identification the potential channels for transferring modern technology from abroad (import-based relations, including investment imports and FDI). Moreover, the rapid development of information technology (including the Internet) that we have been able to observe particularly in the recent years suggests that the research should not miss factors reflecting the use of the state-of-the-art technologies.

References

Aghion P., Howitt P., 1998, *Endogenous Growth Theory*, MIT Press, Cambridge MA.

Aitken B., Harrison A., 1993, *Do Domestically-Owned Firms Benefit from Foreign Direct Investment. Evidence from Panel Data, International Monetary Found*; (mimeo).

Allen R. G. D., 1975, *Teoria makroekonomiczna. Ujęcie matematyczne* (*Macroeconomics theory. Mathematical framework*), PWN, Warsaw.

Antonelli C., 1994, *Technological districts localized spillovers and productivity growth. The Italian evidence on technological externalities in the core regions*, „International Review of Applied Economics", vol. 8, pp. 18–30.

Arrow K.J., 1962, The *Economic Implications of Learning by Doing*, „Review of Economic Studies", vol. 29, pp. 155–173.

Baily, M. N., 1981, *Productivity and the Services of Capital and Labor*, „Brookings Papers of Economy Activity", No. 1, pp. 1–50.

Barro R. J., Sala -i-Martin X., 1995, *Economic Growth*, Mc Graw-Hill Inc., New York etc.

Bernstein J.I., Mohnen P., 1998, International *R&D spillovers between U.S. and Japanese R&D intensive sectors*, "Journal of International Economics", vol. 44, pp. 315–338.

Białoń L., 1974, *Poziom techniki i postęp techniczny w polskim przemyśle elektromaszynowym w latach 1960–1970* (*Level of technology and technical progress in the Polish electro-machinery industry in 1960-1970*), „Gospodarka Planowa", no. 3, pp. 210–220.

Białoń L., Obrębski T. (red.), 1989, *Nauka i technika w rozwoju społeczno-gospodarczym* (*Science and technology in social-economic development*), PWN, Warsaw.

Blaug M., 1994, *Teoria ekonomii. Ujęcie retrospektywne*, (*Economic Theory in Retrospect*), PWN, Warsaw.

Blomström M., 1986, Foreign *investment and productive efficiency: the case of Mexico*, „Journal of Industrial Economics", vol. 15, pp. 97–110.

Borensztein, E. De Gorgio J., Lee J. W., 1998, *How Does Foreign Direct Investment Affect Economc Growth?*, „Journal of International Economics", vol. 45, pp. 115–135.

Bönte W., 2004, *Spillovers from publicly financed business R&D: some empirical evidence from Germany*, „Research Policy", vol. 33, pp. 1635–1655.

Branstetter L., 2001, *Are knowledge spillovers international or international in scope? Microeconometric evidence from the U.S. and Japan*, „Journal of International Economics", vol.53, pp. 53–79.
Bruno M., 1984, *Raw Materials, Profits and the Productivity Slowdown*, „Quarterly Journal of Economics", vol. 49, No. 1, pp. 1–29.
Cameron G., Proudman J., Redding S., 2005, *Technological convergence, R&D, trade and productivity growth*, „European Economic Review", vol. 49, pp. 775–807.
Caves R.E., 1974, *Multinational Firms, Competition and Productivity in Host-Country Industries*, „Economica", vol. 41, pp. 176–193.
Caves D., Christiansen L., Diewert E., 1982, *Multilateral comparison of output, input and productivity using superlative index numbers*, „The Economic Journal", vol. 92, pp. 73–86.
Cincera M., van Pottelsberghe de la Potterie B., 2001, *International R&D Spillovers: A Survey*, „Caniers Economiques de Bruxelles", 169 (1), pp. 1–20.
Clark K.B., Griliches Z., 1982, *Productivity Growth and R&D at the Business Level: Results from the PIMS Database*, NBER Working Paper No. 916.
Coe D.T., Helpman E., 1995, *International R&D Spillovers*, „European Economic Review", vol. 39, pp. 859–887.
Coe D.T., Helpman E., Hoffmaister A.W., 1997, *North-South R&D Spillovers*, „The Economic Journal", vol. 107, pp. 134–149.
Coe D.T., Hoffmaister A.W., 1999, *Are There International R&D Spillovers Among Randomly Matched Trade Patterns? A Response to Keller*, IMF Working Paper No. 18, Washington DC.
Collecchia A., Schreyer P., 2002, *ICT Investment and Economic Growth in the 1990s: Is the United States a Unique Case*, „Review of Economics Dynamics", vol. 5, pp. 408-442.
Cieślak M., 1976, *Modele zapotrzebowania na kadry kwalifikowane,(Models of qualified personel demand)* , PWN, Warsaw.
Crispolti V., Marconi D., 2005, *Technology Transfer and Economic Growth in Developing Countries: an Econometric Ananlysis*, "Working Paper", no. 564, Banca D'Italia.
Czerwiński Z., 1982, *Modelowanie i planowanie wzrostu gospodarki narodowej,* (*Modelling and planning of national economic growth*), PWE, Warsaw.
Drucker P. E., 1992, *Innowacja i przedsiębiorczość. Praktyka i zasady* (*Innovation and initiative. The practice and principals*), PWE, Warsaw.
Englander A.S., 1988, *Research extension and U.S. agricultural productivity: a statistical decomposition analysis*, in: Capalbo S.M., Antle J.M. eds., *Agricultural Productivity Measurement and Explanation,* Washington Presources for the Future.
Fagerberg J., 1987, *A Technology Gap Approach to Why Growth Paths Differ*, „Research Policy", vol.16, pp. 87–99.

Felbur S., 1997, Tendencje postępu technicznego w Polsce (*The technical progress tendency in Poland*), IRiSS, no. 52, Warsaw
Fiedor B., 1986, *Neoklasyczna teoria postępu technicznego. Próba systematyzacji i krytycznej analizy*, (*The neoclassical theory of technical progress. Attempt of systematization and critical analysis*) Wydawnictwo AE, Wrocław.
Glaser J.L., 1993, *Multifactor productivity in the utility services industries*, Monthly Labor Review, May, pp. 34–49.
Gomułka S. , 1998, *Teoria innowacji i wzrostu gospodarczego* (*Innovation theory and economic growth*), Biblioteka CASE, Warsaw.
Goto A., Suzuki K., 1989, *R&D Capital, Rate of Return of R&D Investment and Spillover of R&D in Japanese Manufacturing Firms*, „Review of Economics and Statistics", vol.71, pp. 555–564.
Griliches Z., 1957, *Hybrid Corn: An Exploration in the Economics of Technological Change*, „Econometrica", vol.25, pp. 501–522.
Griliches Z., 1979, Issues *in assessing the contribution of R&D to productivity growth*, „The Bell Journal of Economics", vol. 10, pp. 92–116.
Griliches Z., 1980, R*&D and Productivity Slowdown*, „American Economic Review", vol. 76, pp. 141–154.
Griliches Z., 1981, *A Market Value, R&D and Patents*, „Economic Letters", vol.7, pp. 183–187.
Griliches Z., 1986, *Productivity R&D and basic research at the firm level in the 1970s*, „American Economic Review", vol. 76, pp. 141–154.
Griliches Z., 1995, *Firm productivity in Israeli industry, 1979–1988*, „Journal of Econometrics", vol. 65, pp. 175–203.
Griliches Z., Jorgenson D., 1967, *The Explanation of Productivity Change*, „Review of Economic Studies", July, pp. 349–383.
Griliches Z, Mairesse J., 1983, *Comparing Productivity Growth. An Exploration of French and U.S. Industrial and Firm Data*, „European Economic Review", vol. 21, pp. 89–119.
Griliches Z., Lichtenberg F., 1984, *Interindustry technology flows and productivity growth: a reexamination*, „Review of Economics and Statistics", vol. 64, pp. 324–329.
Gullickson W., 1995, *Measurement of Productivity Growth in U.S. Manufacturing*, „Monthly Labor Review", July, pp. 13–28.
Hanel P., 2000, R&D, *Interindustry and International Technology Spillovers and the Total Factor Productivity Growth of Manufacturing Industries in Canada, 1974–1989*, „Economic Systems Research", vol. 12, pp. 345–361.
Hejazi W., Safarian E., 1999, *Trade, Foreign Direct Investment and R&D Spillovers*, „Journal of International Business Studies", vol. 30, p.491–511.
Hulten Ch., 1978, *Growth Accounting with Intermediate Inputs*, „Review of Economic Studies", vol. 45, October, pp. 511–518.

Jablonski M., 1995, *Multifactor Productivity: Cotton and synthetic broad woven fabrics*, „Monthly Labour Review", July, pp. 29–38.
Jaffe A.B., 1986, *Technological Opportunity and Spillovers of R&D: Evidence from Firms' Patents, Profits and Market Value*, „American Economic Review", vol. 76, pp. 984–1001.
Jaffe A., Trajtenberg M., 1999, *International Knowledge Flows: Evidence from Patent Citations*, „Economics of Innovation and New Technology", vol. 8, p. 105–136.
Jorgenson D.W., 2001, *Information Technology and the U.S. Economy*; „American Economic Review", vol. 90, pp. 1–32.
Jorgenson D., Gollop F., Fraumeni B., 1987, *Productivity and U.S. economic growth*, Harvard University Press, Cambridge, MA.
Jorgenson D.W., Dougherty C., 1996, *International Comparisons of the Sources of Economic Growth*, „American Economic Review", vol. 86, pp. 25–29.
Juszczak-Szumacher G., 1996, *Makroekonometryczna analiza procesu produkcyjnego* (*Macroeconometrics analysis of the production process*), Wydawnictwo UŁ, Łódź.
Juszczak-Szumacher G., 2000, *Pomiar wielkości nieobserwowalnych przez analizę regresji. Postęp techniczny i efektywność procesu produkcyjnego*, (*Measurement of unobservable value by regression analysis. Technical progress and the efficiency of production process*) [in:] Welfe A. (ed.), *Gospodarka polska w okresie transformacji* (*The economy in the transformation process*), PWE, Warsaw.
Kaldor N., 1957, *A Model of Economic Growth,* „Economic Journal", vol. 67, pp. 591–624.
Kalecki M., 1963, *Zarys teorii wzrostu gospodarczego* (*Draft of economic growth theory*), PWN, Warsaw.
Kasperkiewicz W., 1989, *Systemy funkcjonowania gospodarki a innowacje* (*Systems of economy functioning and innovation*) Acta Universitatis Lodziensis, Łódź.
Keller W., 1998, *Are International R&D Spillovers trade-related? Analysing Spillovers among randomly matched trade partners*, „European Economic Review", vol.42, pp. 1469–1481.
Kern P., Wilder P., 1997, *Multifactor Productivity: Refrigeration and Heating equipment industry*, „Monthly Labor Review", May, pp. 6–13.
Kim J., Lau L. J., 1994, The *Sources of Economic Growth of the East Asian Newly Industrialized Countries*, „Journal of Japanese and International Economics", vol. 8, pp. 235–271.
Kokko A., 1996, Productivity *spillovers from competition between local firms and foreign affiliates*, „Journal of International Development", vol.3, pp. 53–63.
Krajewski S., 1985, *Procesy innowacyjne w przemyśle* (*Innovation process in industry*), Książka i Wiedza, Warsaw

Król H., 1974, *Technika, praca i kwalifikacje* (*Technology, work and qualifications*), Książka i Wiedza, Warsaw.
Kudrycka I., 1990, *Możliwości statystycznej oceny postępu naukowo-technicznego i dyfuzji nowoczesnej techniki* (*Possibilities of the statistical assessment of scientific and technological progress and of modern technology diffusion*), „Studia i Prace ZBSE GUS i PAN", 188, Warsaw.
Kunze K., Jablonski M., Klarquist V., 1995, *BLS Modernizes Industry Labor Productivity Program*, „Monthly Labor Review", pp. 3-12.
Leeuwen G. Van, Weil Van der H., 2003, *Innovation and Productivity*, CAED Conference, London
Lichtenberg F., 1984, The *relationship between federal contact R&D and company R&D*; „American Economic Association Paper and Proceedings", vol. 74, pp. 73–78.
Lichtenberg F.R., Van Pottelsberghe de la Potterie B., 1996, *International R&D Spillovers: A Re-examination*; NBER Working Paper No. 5668.
Lichtenberg F.R., Van Pottelsberghe de la Potterie B., 1998, *International R&D Spillovers: A Comment*; „European Economic Review", vol.42, pp. 1483–1491.
Lichtenberg F.R., Van Pottelsberghe de la Potterie B., 2001, *Does Foreign Direct Investment Transfer Technology Across Borders*, „The Review of Economics and Statistics", vol. 83, pp. 490–497.
Los B., Verspagen B., 2000, *R&D Spillovers and Productivity: Evidence from U.S. Manufacturing Microdata*, „Empirical Economics", vol. 25, pp. 127–148.
Lucas R. E., 1988, *On the Mechanism of Economics Development*, „Journal of Monetary Economics", July, pp. 3–42.
MacGarvie M., 2005, *The determinants of international knowledge diffusion as measured by patent citations*, „Economic Letters", vol. 87, pp. 121–126.
Mankiw N. G., Romer D., Weil N., 1992, *A Contribution to the Empirics of Economic Growth*, „Quarterly Journal of Economics", May, pp. 407–434.
Mansfield E., 1968, *Industrial Research and Technological Innovation*, New York, NY:WW Norton.
Maudos J., Pastor J.M., Serrano L., 1999, *Total factor productivity measurement and human capital in OECD countries*, „Economic Letters", vol.63, pp. 39–64.
Meister Ch., Verspagen B., 2004, *European Productivity Gaps: Is R&D the solution*, Centre for Technology, Innovation and Culture, Working Paper 01/04;
Mikosik S., 1993, *Teoria rozwoju gospodarczego Josepha A. Shumpetera* (*Joseph A. Schumeter's theory of economic development*), PWN, Warsaw.
Mohnen P., 1992, *The relationship between R&D and productivity growth in Canada and other industrialized countries*, Ottawa, Minister of Supply and Services Canada.
Nadiri M.I., 1970, *Some Approaches to the Theory of Total Factor Productivity: A Survey*, „Journal of Economic Literature", pp. 1137–1176.

Nadiri M.I., 1993. *Innovations and Technological Spillovers*, NBER Working Paper No 4423.

Nordhaus W. D., 1976, *Innowacje, wzrost i dobrobyt. Postęp techniczny w ujęciu teoretycznym* (*Innovation, growth and welfare. Technical progress in the theoretical framework*), PWN, Warsaw.

Norsworthy J.R., Jang S.L., 1992, *Empirical Measurement and Analysis of Productivity and Technological Change. Applications in High-Technology and Service Industries*, Wydawnictwo North-Holland.

Park W.G., 1993, *International spillovers of R&D investment and OECD economic growth*, Paper presented at the Western Economic Association meetings, Lake Tahoe.

Pawłowski Z., 1976, *Ekonometryczna analiza procesu produkcyjnego* (*Econometrics analysis of production process*), PWN, Warsaw.

Phelps E.S., 1966, *Model of Technical Progress and Golden Rule of Research*, „Review of Economic Studiem", April, pp. 133–145.

Romer P. M., 1990, *Endogenous Technical Change*, „Journal of Political Economy", October, pp. 71–102.

Romer P. M., 1993, Idea *Gaps and Object gaps in Economic Development*, „Journal of Monetary Economics", vol. 32, pp. 543–573.

Scherer F.M., 1982, *Interindustry technology flows and productivity growth*; „Review of Economics and Statistics", vol.64, pp. 627–634.

Schmookler J., 1966, *Invention and Economic Growth*, Cambridge MA Harvard University Press.

Shell K., 1966, *Toward a Theory of Inventive Activity and Capital Accumulation*, „American Economic Review", May, pp. 62–68.

Sjöholm F., 1997, *Technology gap, competition and spillovers from direct foreign investment: Evidence from establishment data*; Working Paper No. 212, Stockholm School of Economics.

Soete L., Verspagen B., 1993, *Convergence and divergence in growth and technical change: an empirical investigation*, Paper presented at the AEA Conference in Manheim.

Solow R.M., 1957, *Technological Change and the Aggregate Production Function*, „Review of Economics and Statistics", pp. 312–320.

Solow R. M., 1967, *Teoria kapitału i stopa przychodu* (*Capital theory and rate of return*), PWN, Warsaw.

Stanek K., 1990, *Wzrost gospodarczy w warunkach postępu technicznego i zmian strukturalnych* (*Economic growth with technical progress and structural change*), PWN, Warsaw.

Sterlacchini A., 1989, *R&D, innovations and total factor productivity growth in British manufacturing*, „Applied Economics", 1989, vol.21, pp. 1549–1562.

Sudit E.F., 1995, *Productivity measurement in industrial operations*, „European Journal of Operational Research", vol. 85, pp. 435–453.

Sveikauskas L., 1981, *Technological Inputs and Multifactor Productivity Growth*, „Review of Economics and Statistics", vol. 68, pp. 275–282.
Sztaudynger J.J., 2003, *Modyfikacje funkcji produkcji i wydajności pracy z zastosowaniami* (*Modifications of production function and labour productivity with applications*) , Wydawnictwo Uniwersytetu Łódzkiego, Łódź.
Świeczewska I., 2004, Postęp techniczny w Polsce. Ujecie sektorowe (*Technical progress in Poland. The sectoral view*). Rozprawa Doktorska, Uniwersytet Łódzki.
Świtalski W., 2005, *Innowacje i konkurencyjność* (*Innovation and competitiveness*), Wydawnictwo Uniwersytetu Warszawskiego, Warsaw.
Terleckyj N.E., 1974, *Effects on R&D on the Productivity Growth of Industries: An Exploratory Study*, National Planning Association, Washington D.C..
Tinbergen J., 1959, *Selected Papers*, North-Holland, Amsterdam.
Tokarski T., 1998, *Postęp techniczny a wzrost gospodarczy w modelach Solowa i Lucasa* (*Technical progress and economic growth in Solow and Lucas models*), „Ekonomista", no. 2–3, pp. 271–291.
Tokarski T., 2001a, *Modele wzrostu endogenicznego* (*Models of endogenous growth*), [in:] Welfe W. ed., *Ekonometryczny model wzrostu gospodarczego* (*Econometric model of economic growth*), Wydawnictwo UŁ, Łódź.
Tokarski T., 2001b, Determinanty wzrostu gospodarczego w warunkach stałych efektów skali (*Determinants of economic growth in the constant scale effects*), Katedra Ekonomii UŁ, Łódź
Tomaszewicz Ł., 1994, *Metody analizy input-output* (*Methods of input-output analysis*), PWE, Warsaw.
Uzawa H., 1965, *Optimum Technical Change in an Aggregative Model of Economic Growth*, „International Economic Review", January, pp. 18–31.
Van Leeuwen G., Van der Wiel 2003, *Spillovers Effects of ICT*, CBP Report, no..3, pp. 24–40.
Verspagen B., 1997, *Measuring Intersectoral Technology Spillovers: Estimates from the European and US Patent Office Databases*, „Economic Systems Research", Vol. 9, No 1, pp. 47–65.
Welfe W., 1966, *Indeksy produkcji* (*Indices of production*), PWN, Warsaw.
Welfe W., 1983, *Koncepcja ekonometrycznego modelu gospodarki narodowej W-5*, (*A concept of an econometric model for a national economy W-5*), Problem resortowy R.III.10, (mimeo), Łódź.
Welfe W. 1992, *Ekonometryczne modele gospodarki narodowej Polski* (*Econometric models of the Polish economy*), PWE, Warsaw.
Welfe W., 2000, *Empiryczne modele wzrostu gospodarczego* (*Empirical models of economic growth*), „Ekonomista", no. 4, pp. 483–498.
Welfe W. (red.), 2001, *Ekonometryczny model wzrostu gospodarczego* (*Econometric model of economic growth*), Wydawnictwo UŁ, Łódź.

Welfe W., 2003, *Łączna produktywność czynników produkcji a postęp techniczny* (*Total factor productivity and technical progress*), „Studia Ekonomiczne", no. 1–2, pp. 99–115.
Wilder P.S., Ziaul Z.A., 2001, *Multifactor productivity trends in manufacturing industries*, 1987–96, „Monthly Labor Review", May, pp. 3–12.
Wolff E.N., 1997, *Industrial Composition, Interindustry Effects and the U.S. Productivity Slowdown*, "Review of Economics and Statistics", vol. 65, pp. 268-277.
Wolff E.N., Nadiri M.I., 1993, *Spillover Effects, Linkage Structure, and Research and Development*; „Structural Change and Economic Dynamics", vol.4, pp. 315–331.
Xu B., Wang, J., 1999, *Capital Goods Trade and R&D Spillovers in OECD*, „Canadian Journal of Economics", vol. 32, p.1258–1274.
Xu B., Wang J., 2000, *Trade, FDI and Internationa Technology Diffusion*, „Journal of Economic Integration", vol. 15, pp. 585–601.
Zheng S., 2002, *Deriving Industry Multifactor Productivity from I-O System*, (nimeo).
Zienkowski L. 2003, *Wiedza a wzrost gospodarczy* (*Knowledge and economic growth*), Scholar, Warsaw.
Żółtowska E., 1997, *Funkcja produkcji. Teoria, estymacja, zastosowania* (*Porducttion function. Theory, estimation and applications*), Wydawnictwo UŁ, Łódź

Chapter 5

Human capital and economic development

5.1. Introduction

Apart from quantitative factors - such as fixed assets and labour force - a long-term socio-economic development is mainly determined by qualitative factors, specifically technological progress, human capital and institutional conditions (Aghion, Howitt [1999], Siwiński [2005]). In addition, many economists tend to believe that human capital should be viewed as a crucial factor because the capacity of a given economy to generate technological progress and use the existent physical capital effectively radically depends upon this production factor (Lucas [1988], Romer [1990a], [2000], Redding [1996], Bucci [2003]).

Human capital in its broad sense is meant as all psycho-physical features of a given individual, such as inborn abilities, personal knowledge, education, job-related skills and experience, health, personal culture, socio-economic activity, philosophy of life, etc., that affect – either directly or indirectly – labour productivity, and are inseparable from a person, being the natural carrier of these values (Shultz [1961], Laroche, Merette, Rugger [1999], The Well Being ... [2001]). In the narrow terms, human capital is usually associated with individual's education. In order to quantify the impact of human capital on economic growth adequate and aggregate measures of its stock must be available, as well as appropriate analytical methods. On the theoretical grounds, the fact that human capital significantly influences labour productivity seems indisputable. Still, verification of this hypothesis meets various predicaments, and outcomes of empirical research do not always confirm that human capital is a statistically significant factor of growth (Bils, Klenow [2000]). Many a time opinions are expressed that the cause-effect link between human capital and growth runs from the latter to the former, and not the other way round. However, the most up-to-date investigations (Hanushek, Kimko [2000], Asteriou, Agiomirgianakis [2001]) seem to confirm that the main direction of the interlink between these two runs from human capital to growth, which does not exclude a role of economic growth for the accumulation of human capital.

5.2 Measuring human capital

5.2.1. Classification of methods. The cost-based approach

Although the concept of human capital is well set in the theory of endogenous growth, its empirical implementation is not easy because a universally accepted measure of human capital does not exist. Most empirical investigations utilize the narrow concept of human capital that associates it with education, but even so, there are a few alternative indicators in this field. The choice of a specific measure is usually dictated by data availability and less often by essential considerations, which adversely affects the quality of estimates and sometimes even leads to outcomes that are contradictory to theoretical postulates. This provoked several researchers to formulate a sad comment about a clear discrepancy between the theoretical models that explicitly take into account human capital and their empirical applications (Temple [1999a], or Borghans, Green, Mayhew [2001]).

There are three basic methodological approaches towards measuring human capital (Le, Gibson [2003]):

a) a cost-based approach, also known as a retrospective method,
b) an educational stock-based approach,
c) an income-based approach, also known as a prospective method.

In the cost-based approach, the stock of human capital in period T equals the streams of discounted outlays spent on its accumulation until the present time and depreciated using a depreciation rate:

$$hc_T = \sum_{t=0}^{T} C_t (1+i-d)^{T-t} \qquad (5.1)$$

where:

hc_T - human capital per worker computed by the cost-based method,
C – costs of investing in human capital including lost benefits (giving up a job to study),
i – discount factor,
d – depreciation factor.

Basic controversies over the use of the cost-based method are due to problems with defining the costs of investment in human capital. And so, Kendrick [1976] numbers among such costs all expenditure connected with upbringing a child to the age of 14, as well as all the spending – private and public – on health care, safety, education and self-improvement, and finally alternative costs. However, only 50% of individual expenses related to health care and safety should be classified as components of human capital. In turn, Machlup [1984] negates the necessity of putting all the costs of child upbringing in the aggregate measure of human capital. Besides, Kendrick leaves out outlays spent on R&D, unlike Eisner [1985].

All the above, as well as some other objections against the cost-based method, such as its ignoring the costs' effectiveness, omitting the problems of lags between investment outlays and stocks (Jorgenson, Fraumeni [1989]), or finally – arbitrary selection of the discount factors make it that the full cost-based method as given by formula (5.1) does not have too many advocates and its empirical applications are rather scarce. Additionally, a vast majority of empirical applications using the cost-based approach to investigate economic growth and human capital employ only some selected costs, mostly outlays on health and education, or their shares in GDP (Welfe [2005]). These measures are based on streams rather than stocks, which should be viewed as a basic shortcoming of such a simplification.

5.2.2. The educational stock-based approach

In empirical research building on international data, the educational stock-based approach, or its modifications, is usually used. First investigations of this type (Romer [1990]) utilized the adult literacy rates:

$$HL = \frac{LL}{L} \tag{5.2}$$

where:

LL – literate adult population of a given country,

L – adult population of a given country.

Even though the measure reflects the stock of human capital, it only accounts for the basic skills acquired in the initial stages of education, thus ignoring mental skills of higher order: logical and analytical reasoning, scientific or technical knowledge. Consequently, it can serve as a useful approximation of human capital only in developing countries, where illiteracy continues to be a major obstacle to economic growth.

Another measure that various empirical analyses employ (Barro [1991], Mankiw, Romer, Weil [1992], Levine, Renelt [1992]) is enrolment rates:

$$WSK_i = \frac{S_i}{L_i} \tag{5.3}$$

where:

S_i - number of people learning at the i-th educational level (elementary, secondary, tertiary),

L_i - number of people whose age corresponds to the i-th level of education.

The gross and net enrolment rates can be distinguished. The first measures the total number of students at a given educational level in relation to the number of persons at the appropriate age, whereas the other only indicates the number of students in an age group corresponding to the formal schooling age at a given educational level.

One can easily note that the above-mentioned indicators by definition cannot measure stocks of human capital, as they are simply streams. Besides, enrolment rates computed for a given year do not translate into the current stock of human capital, because the latter only indirectly depends on the lagged values of the enrolment rates. At the same time, such lags can be very long and variable, depending on the point of time at which people decide to give up studying to start a job. Moreover, the indicators do not take into account changes in the effective labour force, resulting from outflows caused by worker retirement or disability. In brief, the enrolment rates are imperfect measures of human capital (Hanushek, Kimko [2000]).

Criticism expressed against adult literacy rates or enrolment rates made researchers propose some alternative measures that would explicitly address the fact that human capital embodied in labour force is a function of cumulated educational outlays. In these proposals, an average number of years spent on education per worker stands for a proxy of human capital. There are three ways of measuring human capital using the educational stock-based method. These are the perpetual inventory method, the projection method, and the attainment census method. All of them cumulate streams of investment outlays on education as reflected by enrolment rates into a stock of human capital, and their dissimilarities stem from the diverse ways of processing the streams. The proposed measures of human capital were employed in many empirical investigations, such as Barro, Sala-i-Martin [1995], Barro [1997], [2001], Benhabib, Spiegel [1994], Gundlach [1995], Islam [1995], Krueger, Lindahl [2001], O'Neill [1995], Temple [1999b], Welfe [2001] to mention just a few.

The first of the three methods – the perpetual inventory method (Lau, Jamison, Louat [1991] and Nehru, Swanson, Dubey [1995]) – generates a stock of human capital in period T in line with the following formula:

$$HCS_T = \sum_{t=T-A_{\max}+A_0}^{T-A_{\min}+A_0} \sum_i WSK_{i,t+i-1}(1-r_i-d)p_{i,t+i-1} \tag{5.4}$$

where:

$WSK_{i,t}$ - gross enrolment ratio for the i-th level of education at time t,

$A_{\max}$ - retirement age,

$A_{\min}$ - minimal legal age of starting a job,

r_i - share of students repeating a grade at a given educational level (constant in time),

d – share of dropouts (constant in time, for all educational levels),

$p_{i,t}$ - probability that a person with the i-th education will survive from period t to period T (the probability is computed based on the survival tables, assuming that the death rates are unrelated to education).

The value of human capital per worker, hcs_T, is obtained by dividing (5.4) by the number of the economically active persons (LS_T):

$$hcs_T = \frac{HCS_T}{LS_T} \tag{5.5}$$

The second method, i.e. the projection method, takes advantage of a bridge equation where average educational attainment is the following function of enrolment rates:

$$hpr_T = \alpha_0 + \alpha_1 WSKP_{T-15} + \alpha_2 WSKS_{T-10} + \alpha_3 WSKW_{T-5} + \varepsilon_t \tag{5.6}$$

where:

hpr - average educational attainment,

WSKP, *WSKS*, *WSKW* – enrolment rates for elementary, secondary and tertiary education, respectively,

$\alpha_0, \alpha_1, \alpha_2, \alpha_3$ - structural parameters,

ε_t - disturbance term.

Estimation of equation (5.6) based on a given data sample makes it possible to compute predicted educational attainment ($h\hat{p}r$) for periods and/or countries when explicit data are not available, using only information about the enrolment rates. The projection method proposed by Kyriacou [1991] served to construct human capital databases for most countries in the world on the grounds of a 42-country sample. It is quite obvious that for a projection to be reliable a hardly realistic assumption about stability of the relationship between educational attainment and enrolment rates in both time and space has to be made. It is worth mentioning that Kyriacou's results raise many doubts (low coefficients of determination, autocorrelation of error terms). In the light of the aforementioned remarks, it should come as a surprise that estimates of human capital offered by the projection method constituted a basis for numerous investigations (e.g. Benhabib, Spiegela [1994]). It seems that poor data quality was the reason for which they failed to confirm a favourable influence of human capital on economic growth (see de la Fuente, Domenech [2000]).

The last of the educational attainment methods – the attainment census method – makes use of information contained in censuses and surveys. With this type of data, it is possible to arrive at the level of human capital directly, by employing the following formula:

$$hac_T = \sum_{i=\min}^{\max} [n_{iT} (\sum_{i=\min}^{\max} D_{iT})] \qquad (5.7)$$

where:

i – educational level (e.g. elementary, secondary, tertiary),

n_i - fraction of the working population with the i–th education level,

D_i - official length of education at the i-th educational level.

In the method under consideration, people who gave up studying after entering a given educational level are assigned a half of the total time at that level if information that is more detailed is not available.

All the aforementioned measures of human capital based on educational attainment can be fundamentally criticized for two reasons. Firstly, they assume a linear increase in human capital, proportional to the length of time spent at school. Each year at school is given an equal weight, which implicitly means that labour productivity with respect to education is not subject to the law of diminishing returns. However, empirical research shows (e.g. Psacharopoulos [1994]) that the marginal labour productivity decreases together with increasing education.

Secondly, the measures under consideration do not allow for changes in the quality of educational system, either in time or in space. This means that a year of studying in the USA is tantamount to a year in a Bangladesh school, and that the effective values of tertiary education in Japan 40 years ago and now are equal, if the official length of such education has not changed.

5.2.3. The income-based approach

In early one-sector models of growth, the only production factors were fixed capital and labour meant as the total working population. The concept of human capital did not exist in such models, because labour force was regarded as fully homogenous and supplementary. However, even the models' constructors were aware of these shortcomings (see e.g. Solow [1957]), which led to a modification of the concept of labour force to account for its heterogeneity.

The income-based method rests on a crucial assumption that different earnings of individuals reflect their marginal productivities. Hence, wages can serve as weights in building the aggregate measure of human capital. This way, we can cope with the first of the abovementioned critical arguments.

The construction of an aggregate indicator of human capital can be based on the microeconomic Mincer wage equation (Mincer [1970]). This equation makes wages dependent upon education and job experience:

$$\ln W_i = \alpha_0 + \alpha_1 S_i + \alpha_2 X_i + \alpha_3 X_i^2 + \varepsilon_t , \quad (5.8)$$

where:

W_i - wages of the *i*-th worker,

S_i - education of the *i*-th worker measured by educational attainment,

X_i - job experience of the *i*-th labourer measured by the length (years) of their working years (excluding time devoted to formal education).

An appropriate measure of human capital provided by the above formula is given as:

$$HM = e^{f(s)} N \Leftrightarrow hm = e^{f(s)} \quad (5.9)$$

where:

HM – stock of human capital obtained by means of formula (5.8),

L – working population,

hm – stock of human capital per worker: $hm = HM / L$,

$f(s)$ - function of education effectiveness reflecting the marginal productivity of a worker with *s*-year education with respect to a worker without any education.

In the case of the Mincer wage equation the function of effectiveness takes the following form:

$$f(s) = r \cdot s \quad (5.9a)$$

where:

r – rate of return on education derived from the Mincer wage equation (equal to the estimate at variable *S,* which stands for educational attainment).

However, if one needs to diversify between the rates of returns not only with respect to the length of education, but also with respect to various levels of education, then the augmented Mincer wage equation can be used, which has the following form:

$$\ln W_i = \alpha_0 + \alpha_1 D1_i + \alpha_2 D2_i + \alpha_3 D3_i + \alpha_4 D4_i + \alpha_5 X_i + \alpha_6 X_i^2 + \varepsilon_t, \tag{5.10}$$

where:

D1, D2, D3 – dummies measuring educational level attained by the *i*-th worker (respectively: elementary, secondary, and tertiary).

The effectiveness function takes the following form:

$$f(s) = \sum_{lev=\min}^{\max} r_{lev} \cdot s_{lev}, \tag{5.11}$$

whereas total and per worker human capital stocks, respectively, are given as

$$HML = e^{\sum_{lev=\min}^{\max} r_{lev} \cdot s_{lev}} \cdot L_{lev} \tag{5.11a}$$

$$hml = e^{\sum_{lev=\min}^{\max} r_{lev} \cdot s_{lev}} \tag{5.12}$$

where:

r_{lev} - rate of return for a given educational level (elementary, secondary, tertiary), obtained from the augmented Mincer wage equation,

s_{lev} - average educational attainment (length of education at a given educational level), in years,

L_{lev} - working population with completed education at a given level.

An alternative measure of human capital, also based on the microeconomic Mincer wage equation, was proposed by Mulligan and Sala-i-Martin [1997]. The indicator they used was a quotient of an average wage in a given economy (*WAV*) over an average wage earned by the uneducated worker:

$$hw = \frac{WAV}{WNO} \tag{5.13}$$

If appropriate information on variable *WNO*[1] is missing, then it is derived by means of the Mincer wage equation.

The measure of human capital as given by formula (5.13) rests on the correct observation that average wages in an economy are determined not only by the overall educational attainment of the society but also by the labour-capital ratio, and by the modernity of physical capital employed in the production process. On the other hand, however, this implicitly assumes that human capital embodied in an unskilled labourer is constant in both time and space. Besides, wage relations are formed in a clearing process, as a resultant of demand and supply adjustments. In the well-educated societies typified by a comparatively

[1] In developed countries the share of people without any education is negligible, and these are only elderly people who did not attend even the elementary school in their childhood. Mandatory education - even if available - was not always complied with before World War II, and during the war it was even abolished in many occupied countries.

low supply of ill-skilled workers, this may lead to a relative increase in the wages of such workers, and consequently – to a lower aggregate measure of human capital. The minimum legal wages also matter for the measure under consideration.

Human capital measures based on the Mincer wage equation were employed in manifold researches, to mention those by Klenow, Rodriguez [1997], Hall, Jones [1999], Jovanovic, Rob [1999], or Bils, Klenow [2000].

An analogous indicator, frequently used in empirical investigations, is based on the same theoretical premises as (5.13) but it utilizes exclusively macroeconomic data (Hers [1998] or Welfe et al. [2002]):

$$HKLZ = \sum_{i=\min}^{\max}(W_i / W_{\min}) \cdot L_i \tag{5.14}$$

where:

W_i - average wages of workers with the i-th educational level,

$W_{\min}$ - average wages of workers with the lowest (none or elementary) educational level, L_i - number of workers with the i-th educational level.

Measure (5.14) is highly aggregated. It takes into account only decomposition of the labour force by education. Denison [1966] proposed an aggregate indicator of labour force that allowed for its diversification not only by education but also by gender, age and time worked. Having data on the labour force distribution with respect to the aforementioned features, we can assign appropriate weights to each group of workers and compute the aggregate measure of human capital in line with the following formula:

$$H_t = \sum_{a=\min}^{\max}\sum_{b=\min}^{\max}\sum_{c=\min}^{\max} w_{abcK_t} \cdot L_{abcK_t} + \sum_{a=\min}^{\max}\sum_{b=\min}^{\max}\sum_{c=\min}^{\max} w_{abcM_t} \cdot L_{abcM_t}, \tag{5.15}$$

where:

H – labour force including human capital,

a, b, c – summation intervals for age, education, and time worked, respectively,

K – female,

M – male,

L_{abcK}, L_{abcM} - number of workers in the specific group,

w_{abcK}, w_{abcM} - wages/weights normalized with respect to the lowest wage,

min, max – the lowest and the highest values in appropriate intervals by a, b, c, K, M.

Jorgenson [1995] goes even further in his disaggregation of factors determining the quality of labour force. Apart from the features reported above, he distinguishes various professional groups as well as occupational status, and runs a branch disaggregation with respect to all these features. The methodology Jorgenson suggested is much the same as that proposed by Denison, the only difference being the fact that formula (5.15) is extended by further summation operators to include individual branches, professions and professional status. Compared with the methodology proposed by Denison, the Jorgenson' measure takes into account changes in both the occupational structure of labour force and the structure of production, as well as their influence on human capital. Bearing in mind quick technological changes that typify the contemporary developed economies, and the consequent dynamics of the structure of labour force and production, the modification introduced by Jorgenson seems fully justified.

Dagum and Slottje [2000] combine the Jorgenson's methodology with a microeconomic estimation of human capital using data on households, which enables them not only to compute an aggregate value of human capital, but also to fix its distribution by various types of households.

One can easily notice that the above methods of estimation of human capital, although methodologically rather straightforward, are very data intensive. Consistent estimation of human capital based on the methods involves either full or representative research into wages, addressing all the aforementioned occupational features and conducted on an annual basis. Few national statistical offices carry out such research with annual frequency. Therefore, it is frequently assumed for pragmatic reasons that weights in equation (5.15) are constant, which enables implementation of the described methods even if only a few or just one (i.e. for one period) adequate observation is available. Still, even under this strong assumption the crucial issue in implementing the methods remains construction of detailed age-specific population tables, and for the purpose of forecasting – explanation of vital statistics of population. It is worth noting that the introduction of population age-structure into the estimates of human capital makes it possible to account implicitly for the effects of learning by doing. Job experience is primarily the function of the years of service, directly related to worker's age.

The reported measures of human capital are mainly employed in econometric analyses of specific countries in specific periods that are based on the time series data. Although the measures exhaustively account for various factors that significantly affect labour productivity/quality, they have been used so far in relatively few empirical analyses. It seems that the main reason for this situation is the already mentioned difficulty in collecting credible data.

5.2.4. Measuring the quality of human capital

Initially, it was attempted to solve problems resulting from the omission of the quality factor in the measures of human capital by introducing into regressions explaining economic growth various proxies as independent variables. The following variables served this purpose in empirical investigations:

a) number of students per teacher (Barro [1991]);
b) share of educational outlays in GDP (Barro, Sala-i-Martin [1995]);
c) expenditure per student, teachers' salaries, length of the school year (Barro, Lee [1996]).

Microeconomic investigations also indicate that it is purposeful to use other measures of the teaching process as proxies of the quality factor in the mentioned regressions. These are:

d) average time spent at school (see e.g. Leuven e al. [2004]);
e) number of students per class (Krueger [2003]);
f) competence of the teaching staff (Angrist, Lavy [2001]).

However, although each of the above measures improves the effectiveness of the teaching process – associated with students' mastering relevant knowledge or skills – none of them, if used separately, is capable of explaining the variability of the process. What is more, from the standpoint of human capital accumulation these variables are streams, and so they describe at best the quality of the current school leavers but not of the total labour force.

Hanushek and Kimko [2000] recommend another approach to constructing a human capital indicator that uses the results of standard tests taken by the school leavers at various educational levels. Using databases developed by several international educational institutions, they built two aggregate measures of human capital normalized with respect to the USA. The measures, computed for 31 countries and 6 years in the period 1965-1991, are properly weighted to account for the age structure of the working population. As a result, the measures represent stocks rather than streams. The authors suggest explaining the variance of their measures using variables that *per se* used to represent human capital in the previous investigations, such as the number of students per one teacher or the share of educational expenditure in GDP. In this way, they endogenize the process responsible for the quality of human capital. Estimates standing at respective explanatory variables describe the effectiveness of individual outlays.

Unlike Haushek and Kimko [2000] – who separated human capital quality from its quantity (the latter understood as educational attainment) - Woessman [2003] combines the two elements into one measure and uses information obtained from the Mincer wage equation:

$$hq_i = e^{\sum_{a=\min}^{\max} r_a \cdot Q_i \cdot S_{ai}} \qquad (5.16)$$

where:

hq_i - human capital by Woessman,

r_a - average for global rate of return on the a-type education (no education, elementary, secondary, tertiary education),

Q_i - aggregate index of the quality of human capital by Hanushek-Kimoko for the i-th country,

S_{ai} - statutory length of education (in years) at the a-th level of education in the i-th country.

The measure proposed by Woessman addresses all key aspects of human capital associated with formal education, from the quantitative factors (formal length of education) through the qualitative factors (the Hanushek-Kimko index) to step changes in the productivity of workers with different educational background (returns to education).

It should be mentioned that the methodology recommended by Hanushek and Kimko, and further developed by Woessman, might face hardly surmountable problems if it were to be applied to time series data to account for human capital changes in a given country in time. The main reason is the lack of comparable data across time to produce appropriate measures. Still, the approach initiated by Hanushek and Kimko marks a promising direction of research, which explicitly addresses the effectiveness of educational outlays.

5.2.5. Measures of the broad concept of human capital

Measures discussed in the previous sections deal with the narrow concept of human capital, associated with education only. However, its broad concept embraces also other psychophysical features of an individual, such as health, personal culture, socio-economic activity or philosophy of life that – directly or indirectly - affect labour productivity. Although the role of such factors is unquestionable, there have been very few attempts so far to incorporate them into a synthetic index of human capital because of difficulty in collecting the appropriate data. It should be mentioned that in empirical investigations the features quoted above – approximated by adequate indicators[2] - are many a time treated as exogenous variables explaining the dynamics of economic growth (e.g. Sala-i-Martin [1997]).

Measures employed in empirical analyses accentuate the role of some selected, crucial components of human capital, predominantly education, health and job experience[3]. Empirical models of growth show that any attempt at including simultaneously several indicators of human capital in one regression fall flat. Hence, the following measure of human capital is suggested, comprising each of the three crucial components of human capital:

$$hf = e^{\sum_{lev=min}^{max} r_{lev} \cdot s_{lev}} \cdot e^{\sum_{i=18}^{65} \frac{\alpha_1 i + \alpha_2 i^2}{\alpha_1 18 + \alpha_2 18^2}} \cdot le \tag{5.17}$$

where:

$e^{\sum_{lev=min}^{max} r_{lev} \cdot s_{lev}}$ - average educational attainment (see (5.12)),

[2] For instance, the health status of a society is most frequently associated with life expectancy at birth, whereas philosophy of life – with the religious denomination.

[3] It is worth noting that job experience can be largely associated with the concept of learning by doing.

$e^{\sum_{i=18}^{65} \frac{\alpha_1 i+\alpha_2 i^2}{\alpha_1 18+\alpha_2 18^2}}$ - job experience approximated by the age structure of labour force[4] (parameters α_1 and α_2 are taken from the Mincer wage equation),

le – life expectancy at birth approximating society's health status.

All the components of measure (5.17) represent stocks and have equal weights.

It is possible to enumerate alternative indicators of human capital. Florczak [2007], motivated by theoretical considerations (see e.g. Świtalski [2000]), augments the traditional measure (5.14) by adding effects related to the popularization of knowledge among the society and constructs a combined variable comprising the average number of library loans. This modification considerably improved the results of his analyses. Milo and Szafrański [2006] quote several proposals of human capital, including an indicator based on employability related to person's education. Uramek [2006] employs alternative taxonomic methods to arrive at a synthetic measure of human capital that comprises five traditional indexes, and introduces it to an empirical model of growth of the Mankiw-Romer-Weil type, confirming its statistical significance to growth.

While deliberating on the definitions, measures and the impact of human capital upon economic growth we need to distinguish between the aggregate level of human capital and its average level per worker (see e.g. Welfe [2001]). The aggregate measure, also known as the effective labour force, is constructed as a product of the working population by a unit measure of human capital. The approach arises from the fact that the aggregate value may exhibit a decline, whereas the unit measure an increase. This situation is caused by changes in total employment, e.g. because of rapidly rising unemployment. This distinction is critical to the credibility of comparisons.

There are two different approaches to the role of aggregate human capital and unit human capital for economic growth. The first approach originates from modifications of the classical models of growth and treats effective labour force (aggregate human capital) as a production factor. Consequently, elasticities of production with respect to "raw" labour force and unit human capital are the same. The other approach, common to neoclassical and endogenous models of growth, treats unit human capital as a separate production factor. Consequently, elasticities of production with respect to labour force and unit human capital may differ (Canton et al. [2005]).

The overview presented in this section leads to the conclusion that despite the role assigned to human capital in the long-term economic growth its fully adequate and universal measures are still missing. It should come as no surprise

---

[4] Job experience is by no means a function of workers' age, which justifies the above approximation.

that the coefficients of correlation between the measures of human capital most frequently used in empirical research are relatively low (see Woessmann [2003]). Consequently, it matters a lot which of them are selected to investigate the relationship between production factors and economic growth (see section 5.4).

Human capital, associated with education, fundamentally determines the so-called social capital (see e.g. Hall, Jones [1999] or Bjoornskov [2006]). Social capital is meant as the effectiveness of functioning of all societies or countries (see Putnam [1993], Sztaudynger [2003]). It is decisive to the quality of the institutional setting of economic activity, and so it determines socio-economic development in the long run.

5.3 Empirical models of growth with human capital

5.3.1. Structural models

The mainstream investigations into the role played by human capital in the socio-economic development of various countries are based on macroeconomic analyses, in which models of growth are of great importance. The existing postulates and theoretical solutions have a far-going bearing on empirical research. In particular, one should distinguish between structural and non-structural approaches towards modelling economic growth with the use of human capital (Canton et al [2005]).

Structural, empirical models are derived from theoretical constructions. They are characterised by consistency with the microeconomic underpinnings. The number of explanatory variables in such models is relatively low, and their selection is strongly supported by economic theories. Depending on the theoretical premises, human capital becomes part of the structural models either as a separate explanatory variable (or in the form of a logarithm / increment in logarithm), or as an interaction variable.

Within the structural approach, five main hypotheses of the role of human capital as growth determinant can be distinguished (Gemmell [1996]). They imply the way of introducing a human capital variable to the empirical models of growth.

According to the oldest hypothesis, human capital is a component of effective labour force (see formula (5.2)) and does not enter a production function as a separate variable. The elasticity of production with respect to human capital is then equal for all components defining the effective labour force.

The second hypothesis treats human capital as an additional production factor, beside physical capital and labour force. As a result, the simplest empirical verification of this hypothesis leads to the construction of an augmented production function of the Cobb-Douglas type. Then the rate of GDP growth directly depends upon the dynamics of human capital.

The third of the hypotheses associates the impact of human capital on growth with technological progress whose formation is crucially influenced by human capital, and with the spillover effects of technical knowledge originated abroad (see e.g. Nelson-Phelps [1966] or Baumol [1986]). Therefore, the dynamics of economic growth can be determined not only by the dynamics of human capital stock but also by its absolute level, which justifies introduction of the two factors into a growth equation. An alternative verifiable hypothesis based on the outlined premises consists in inserting the human capital stock in the initial period of the sample. Parameter standing at this variable informs about the extent to which countries typified by high levels of human capital enjoy, *ceteris paribus*, higher rates of economic growth. In particular, this means that such countries may have high dynamics of growth despite low dynamics of human capital.

Hypothesis four (see Lucas [1988], Romer [1990b]) accentuates the role of human capital as a stimulator of innovation, scale and effectiveness of R&D outlays that in turn generate technological progress. In the light of this hypothesis human capital affects growth directly, by interacting with physical capital and/or labour force, or through total factor productivity. Empirical verification of hypothesis four leads to the construction of models with interaction variables and total factor productivity as a function of human capital. Impacts of human capital upon production are estimated in two stages. First parameters of the production function are estimated (calibrated) and total factor productivity is determined residually. Next parameters standing at the human capital variable are estimated, with the explained variable representing total factor productivity.

According to hypothesis five (see e.g. Manikiw, Romer, Weil [1992]) it is not the absolute level of human capital that influences economic growth in a given country but rather its relative value, measured as the share of investment outlays on human capital in GDP, which justifies the introduction of this variable into the production function.

The aforementioned hypotheses of the channels of human capital influence upon economic growth are rather complementary against each other. Hence, empirical analyses sometimes attempt at simultaneous verification of several, or even all the hypotheses by means of one-equation econometric models (see e.g. Gemmell [1996]).

5.3.2. Non-structural models

Non-structural models – often called *a la* Barro models – utilize panel data and the following general form of regression:

$$d\ln y_{it} = \alpha_{it} + \beta_1 X_{1it} + \beta_2 X_{2it} + \ldots + \beta_k X_{kit} + \varepsilon_{it} \qquad (5.18)$$

where:

$d\ln y_{it}$ - rate of GDP growth (increment in logarithm) per worker in the i-th country in period t,

X_{kit} - value of the k-th explanatory variable (including human capital) in the i-th country in period t (depending on theoretical considerations the explanatory variables are either increments in logarithms, logarithms or levels),
$i = 1, 2, ..., N$ – index marking objects (countries),
$t = 1, 2, ..., T$ – index marking periods (years),
$k = 1, 2, ..., K$ – index marking explanatory variables,
α_{it} - intercept (variable in time and by object),
β_k - structural parameters of the model, (constant in time and by object),
ε_{it} - disturbance term satisfying the Gauss-Markov assumptions (white noise).

Explanatory variables are frequently selected for equation (5.18) in an *ad hoc* manner, without caring too much about the theoretical premises of a given specification. An *implicit* assumption is made that all the explanatory variables are direct factors of production, which can be hardly justified. Most explanatory variables used in empirical models of growth should be rather regarded as determining total factor productivity, and so their impact on growth should be modelled via TFP (see e.g. Hall, Jones [1999] or Miller, Upadhyay [2000]).

Regressions *a la* Barro usually employ peculiar cross-sectional data, which are either averages or average rates of growth of appropriate variables over relatively long periods of several tens of years. Besides, some explanatory variables in equation (5.18) take on values only for the first observation in the sample. Then equation (5.18) boils down to a standard cross-sectional regression:

$$d\_\ln y_i = \alpha + \beta_1 XL_{1i} + \beta_2 XL_{2i} + ... + \beta_k XL_{ki} + \varepsilon_i \qquad (5.19)$$

where:
$d\_\ln y_i$ - annual, average long-term rate of growth of GDP per worker in the i-th country,
XL_{ki} - value of the k-th explanatory variable (including human capital) in the i-th country.

Depending on the theoretical considerations, the explanatory variables are given as their long-term average rates of growth, logarithms of their average long-term values, or finally as their long-term average values.

The reason for the long-term average rates of growth to be used instead of standard panel data arises from the belief that this helps eliminate the impact of the short-term demand adjustment on the results. Notwithstanding, equation (5.19) assumes that TFP effects are constant in time and across the sample countries. This is one of the main objections raised against equation (5.19) (Bassanini, Scarpetta [2001]). The number of observations purposefully reduced to only one per country calls for the inclusion in the estimation sample of many countries whose levels of civilization development are considerably diversified.

Another frequently practised approach uses a few-year (usually 5-year) averages. It is then possible to estimate the intercept, α_i, in equation (5.18) for each country, or even with respect to time, α_{it}, and consequently to account for the lack of homogeneity in the sample. Still, it follows from equation (5.18) that structural parameters measuring the influence of the other production factors, excluding TFP, are constant by object and thus – as Scarpetta et al. [2000] proved – consistent estimators can be obtained only under very unrealistic assumptions[5].

Two methods can be applied to solve this problem. One limits the sample to relatively homogenous countries on *a priori* grounds (see e.g. Gemmell [1996]), or through *a posteriori* elimination of the outliers (see e.g. Temple [1999b]). The other approach, much less common because of data scarcity, employs the method of mean group estimators (see Evans [1997]).

Another practical problem with the implementation of empirical growth models *a la* Barro is due to the multitude of possible explanatory variables, whose statistical significance has been proven in at least one research. The macroeconomic policy makes us raise two questions here. Firstly, which of the possible regressors should be regarded as actual determinants of growth? Secondly, to what extent are estimates standing at a given variable (e.g. at human capital) conditioned by the presence of other variables in the growth regression?

An attempt to answer these questions was made by Levine and Renelt [1992] and then Sala-i-Martin [1997]. They took advantage of the extreme bounds test, elaborated by Leamer [1985], to show a set of variables that are robust to changes in the specifications of an empirical model of growth.

5.3.3. Human capital databases and reasons for different empirical results

Among factors affecting the empirical results, we need to mention the type of measure used to approximate human capital and international databases on human capital applied to empirical research. The following bases are usually quoted as the sources of information in such research: Summers, Heston [1988], [1991], Kyriacou [1991], Lau, Jamison, Louat [1991], Lau, Bhalla, Louat [1991], Barro, Lee [1993], [1996], [2000], Nehru, Swanson, Dubey [1995], de la Fuente, Domenech [2001], and Cohen, Soto [2001].

The databases are generally built with data on the enrolment rates and – although less frequently – educational attainment and they draw on information published in UNESCO's statistical yearbooks. A comparative analysis of such data reveals their relatively poor quality. The measurement errors can lead to not only an understated magnitude and significance of human capital influence upon growth, but in extreme cases also to negative signs of the estimates (see e.g. Temple [1999b], Kruger, Lindahl [2001], or De la Fuente, Domenech [2000]). However, smart transformation procedures allowing for the measurement errors (see e.g. de la Fuente [2004]) provide better results. It is fully justified to use the

[5] The assumptions refer to the similarity of technological development, and to the equallity of natural increase across the sample countries.

word 'procedure' in the plural, because a single universal correction procedure does not exist (see e.g. Temple [1999], Krueger, Lindahl [2001], Fuente, Domenech [2001]).

By way of recapitulation of the sources of variations in the estimates of the human capital influence on economic growth, we can attribute the variations to the following factors:

1. Data sources.
2. The measure of human capital used.
3. Estimation sample, in particular:
   a) data frequency (cross-sectional data, panel data, panel data of 5- and 10-year averages, time series data),
   b) the time span of the sample,
   c) the number of countries in the sample and the sample's homogeneity (all countries in the world, OECD countries, developing countries, etc.).
4. Specification taken in the growth equation, in particular:
   a) number of explanatory variables,
   b) functional form (logarithmic, increments in logarithms),
   c) variability of parameters in time (constancy of parameters in time and by object, fixed effect in objects, fixed effect in time and in objects).
5. Estimation method.
6. Assumptions about the scales of returns, total factor productivity, and mechanisms linking human capital with economic growth.

5.4 Survey of outcomes based on empirical models of growth with human capital

5.4.1 Initial remarks

Most empirical models of growth start with an aggregate production function in one of the below forms:

$$Y_{it} = A_{it} \cdot K_{it}^{\alpha_K} \cdot H_{it}^{\alpha_H} \cdot L_{it}^{\alpha_L} \cdot e^{\varepsilon_{it}} \tag{5.20}$$

or:

$$Y_{it} = A_{it} \cdot K_{it}^{\alpha_K} \cdot e^{\theta \cdot H_{it}} \cdot L_{it}^{\alpha_L} \cdot e^{\varepsilon_{it}} \tag{5.21}$$

where:

Y_{it} - production (GDP) in the *i*-th country in year *t*,

A_{it} - total factor productivity,

K_{it} - fixed assets stock,

H_{it} - average human capital per worker,

L_{it} - number of working persons,

e - base of natural logarithms,
$\alpha_K, \alpha_H, \alpha_L, \theta$ - structural parameters,
ε - error term.

Formulae (5.20) and (5.21) differ with respect to the way the human capital variable is introduced. Function (5.21) is called the Mincer specification.

Functions (5.20) and (5.21) can be transformed into a linear form with respect to parameters using a logarithmic operator, or logarithmic and difference operators together. As a result, the following estimable equations emerge:

a) for logarithms:

$$y_{it} = a_{it} + \alpha_K k_{it} + \alpha_H h_{it} + \alpha_L l_{it} + \varepsilon_{it} \qquad (5.22)$$

$$y_{it} = a_{it} + \alpha_K k_{it} + \theta \cdot H_{it} + \alpha_L l_{it} + \varepsilon_{it} \qquad (5.23)$$

b) for increments in logarithms:

$$\Delta y_{it} = \Delta a_{it} + \alpha_K \Delta k_{it} + \alpha_H \Delta h_{it} + \alpha_L \Delta l_{it} + \Delta\varepsilon_{it} \qquad (5.24)$$

$$\Delta y_{it} = \Delta a_{it} + \alpha_K \Delta k_{it} + \theta \Delta H_{it} + \alpha_L \Delta l_{it} + \Delta\varepsilon_{it} \qquad (5.25)$$

where small letters denote logarithms, Δ stands for increments, and the variables have the same symbols as in formulae (5.20) and (5.21).

Estimation of equations (5.22)-(5.25) is problematic, because total factor productivity is not directly observable and therefore some simplifying assumptions must be made. Depending on specific assumptions, different estimable versions of equations (5.22)-(5.25) are obtained. In the simplest case it can be presumed that the rate of technological progress is constant both in time and by object ($\Delta a_{it} = g$), which leads us to the following versions (5.24)-(5.25):

$$\Delta y_{it} = g + \alpha_K \Delta k_{it} + \alpha_H \Delta h_{it} + \alpha_L \Delta l_{it} + \Delta\varepsilon_{it} \qquad (5.26)$$

$$\Delta y_{it} = g + \alpha_K \Delta k_{it} + \theta \Delta H_{it} + \alpha_L \Delta l_{it} + \Delta\varepsilon_{it} \qquad (5.27)$$

Another consequence of the above assumption is modification of equations (5.22)-(5.23) that have intercepts varying by object, and a trend common to all the objects:

$$y_{it} = a_i + g \cdot T_t + \alpha_K k_{it} + \alpha_H h_{it} + \alpha_L l_{it} + \varepsilon_{it} \qquad (5.28)$$

$$y_{it} = a_i + g \cdot T_t + \alpha_K k_{it} + \theta \cdot H_{it} + \alpha_L l_{it} + \varepsilon_{it} \qquad (5.29)$$

Alternatively, one can assume that the rate of TFP growth is a function of some variables, including human capital:

$$\Delta a_{it} = \gamma_i + \gamma_G G_{it} + \gamma_H H_{it} + \gamma_{GH} H_{it} G_{it} + \gamma_{RD} RD_{it} \qquad (5.30)$$

where:

G_{it} - variable approximating a technological gap between the i-th country and the most technologically advanced country,
RD_{it} - investment outlays on R&D,
$\gamma_i, \gamma_G, \gamma_H, \gamma_{GH}, \gamma_{RD}$ - structural parameters.

If data on the stocks of production factors are not available, then the growth equation can be derived not from the production function, but from information about investment outlays on the production factors. The point of reference is the generalized Solow model, modified in the way proposed by Mankiw, Romer and Weil [1992]. In this model, shares of the production factors in GDP are a function of investment outlays, assuming a long-term steady state, whereas the deviations of these shares from the long-term equilibrium can be approximated using rates of growth of the investment outlays and the initial level of GDP *per capita*. Under a hardly realistic assumption that most countries are close to their steady states, equation (5.22) can be replaced with an equation in which labour productivity per worker is explained using rates of growth of investment outlays on physical and human capital. For a more realistic assumption that countries are not in their steady states an additional variable must be introduced, namely the initial level of labour productivity per worker. These assumptions translate into alternative specifications of the growth equations.

Under the first assumption, the standard specification of a steady state is the following:

$$q_{it} = a_i + gT_t + \frac{\alpha_K}{1-\alpha_K-\alpha_H}\ln\frac{SK_{it}}{\delta+g+n_{it}} + \frac{\alpha_H}{1-\alpha_K-\alpha_H}\ln\frac{SH_{it}}{\delta+g+n_{it}} \tag{5.31}$$

where:

q_{it} - labour productivity per worker (logarithm) in the i-th country in period t,

δ - depreciation rate of physical and human capital (constant in time and by object),

g – rate of growth of technological progress,

SK – share of investment outlays on physical capital in GDP,

SH - share of investment outlays on human capital in GDP,

n - rate of growth of labour force,

α_K, α_H - elasticities of production with respect to physical and human capital, respectively.

In the case, when the long-term steady state is being approached, the convergence equation takes the following form:

$$\Delta q_{it} = g + \beta\left(\frac{\alpha_K}{1-\alpha_K-\alpha_H}\ln\frac{SK_{it}}{\delta+g+n_{it}} + \frac{\alpha_H}{1-\alpha_K-\alpha_H}\ln\frac{SH_{it}}{\delta+g+n_{it}}\right) + \beta(a_i + gT_t) - \beta q_{it} \tag{5.32}$$

Parameter β measures the speed of convergence to the steady state for a known share of investment outlays in GDP.

Formulae (5.20)-(5.32) mark the general, theoretical framework of specifications of aggregate, one-equation models of growth whose various elements can be found in empirical applications. The nomenclature taken in formulae (5.20)-(5.32) will be used to present outcomes of the empirical growth models that will be discussed in the next section. In particular, the capital letter denotes the level of a given variable, the small letter stands for the logarithm of a given variable, and „Δ" – for an increment in a variable. When a relationship between the given explanatory variable and the explained variable is positive and statistically significant, then the explanatory variable is marked with the "+" sign, but a negative relationship gets the "-" sign. However, if the influence of the given explanatory variable is significant but – object-dependent – positive or negative (quality variables), then this explanatory variable is marked with "yes". Finally, "no" is given to variables whose statistical significance was not confirmed.

5.4.2. Human capital in the convergence models

Table 5.1 reports estimation results obtained in the historically earliest research based on the *a la* Barro models of growth. Such models, representing a conglomeration of specifications presented above, generally utilize cross-sectional data. Most of them do not concentrate on explaining GDP growth in the long term or labour productivity growth, but rather on the convergence hypothesis. In these models, in addition to the initial value of GDP *per capita*, one of the regressors is human capital represented by the literacy ratio, enrolment rate or educational attainment. Human capital proxies are inserted in the growth equation in the form of their levels in the initial period of the sample.

Table 5.1. Estimates of early *a la* Barro models of growth

No.	Source	Measure of human capital	Specification / estimates	Other explanatory variables	Source of data	Sample	Estimation method	Notes
1.	Landau [1983]	Average weighted enrolment rates for primary, secondary and tertiary education	$-0{,}0021\, Q_o + 0{,}026\, H_o$ (6,18) (7,64)	Population size (no), share of public consumption in GDP (-), climate zone (yes)		1961-76, cross-sectional data for 96 countries, N=96	OLS	
2.	Landau [1986]	Average weighted enrolment rates for primary, secondary and tertiary education; share of public educational expenditure in GDP	$-0{,}322 Q_o + 0{,}032 H_o + 0{,}059 SK - 0{,}262 FER$ (4,80) (4,87) (1,37) (1,35) *FER* – fertility rate	Share of investment outlays in GDP (no), share of public educational expenditure (no), population size (-), share of public consumption in GDP (-), trend (no), inflation rate (-), dummy for crude oil exporting countries (+), distance from the nearest harbour (-)		1960-80, panel data (averages for four-year periods), 65 countries, N=151	OLS	
3.	Baumol et al. [1989]	Enrolment rates for tertiary education	1) $0{,}622\, Q_o$ (1,72) 2) $-1{,}47\, Q_0 + 1{,}615\, H_o$ (2,47) (5,0)	Share of investment outlays in GDP (+), non-investment public expenditure (-), dummies for African and Latin-American countries (-)	Summers -Heston [1988]	1960-1981, cross-sectional data, 103 countries, N=103	OLS	

Table 5.1. - continued

4.	Azariadis, Drazen [1990]	Literacy rates	1) $-0{,}00034\ Q_0 + 0{,}0122\ H_0$ (9,69) (3,40) 2) $-0{,}0013\ Q_0 + 0{,}0025\ H_0$ (1,46) (3,09) 3) $-0{,}00013\ Q_0 + 0{,}0103\ H_0$ (1,65) (5,24)		Summers-Heston [1988]	1960-80, cross-sectional, 1) N=31, 2) N=40, 3) N=71	OLS	1) countries below 700 USD per capita 2) from 700 to 3500 USD 3) whole sample
5.	Romer [1990b]	Literacy rates	$-0{,}00063\ Q_o + 0{,}01547\ H_o + 0{,}1884\ SK$ (3,59) (1,89) (6,86)	Non-investment public expenditure (-), dummies for African and Latin-American countries (-)	Summers-Heston [1988]	1960-85, cross-sectional, N=94	OLS	
6.	Barro [1991]	Enrolment rates for elementary (*HP*) and secondary (*HS*) education	1) $-0{,}0075\ Q_o + 0{,}0305\ HS_o + 0{,}025\ HP_o$ (6,25) (3,86) (4,46) 2) $-0{,}0077\ Q_0 + 0{,}01\ HS_o + 0{,}0118\ HP_o$ (8,56) (1,15) (2,07) $+ 0{,}064\ SK - 0{,}004\ FER$ (2,00) (3,07)	Non-investment public expenditure (-), index of deviation from PPP (-), number of revolutions and coups (-), number of political assassinations (-), *FER* – fertility rate	Summers-Heston [1988]	1960-85, cross-sectional, N=98	OLS	
7.	Kyriacou [1991]	Educational attainment	$-0{,}009\ Q_o + 0{,}0062\ HS_o$ (2,43) (4,09)		Kyriacou [1991]	1970-85, cross-sectional, N=89	OLS	

It follows from table 5.1 that in order to confirm the convergence hypothesis it in necessary to introduce into the growth equation not only the initial value of GDP but also the initial value of human capital (see variant [3]). This means that countries with lower GDP *per capita* – including the emerging economies – may enjoy higher dynamics of growth vis-à-vis technologically advanced economies, if the level of their human capital is a suitably high. This conclusion holds irrespective of whether human capital is approximated by streams (enrolment rates) or stocks (educational attainment).

Incorporating investment outlays and fertility rate to the growth equation reduces the estimates at human capital, which implicitly testifies to the fact that human capital is adverse for fertility, but favourable for the investment outlays (human capital and physical capital are complementary with respect to each other).

In more up-to-date *a la* Barro models, a simultaneous use of more than one measure of human capital in the growth equation can be observed. In addition to the aforementioned measures, cost indexes represented by the share of investment outlays on education in GDP or health indicators in the form of life expectancy are also used in such equations. Although narrowly defined human capital (associated with education) affects society's health status through positive externalities, it is by no means an adequate proxy of this status. This observation justifies the introduction of life expectancy as an additional variable to the growth equation. Besides, these models accentuate the role of many other explanatory variables approximating institutional, political and economic conditions, including human capital interactions with such variables. Concise results of the discussed research are presented in table 5.2.

The results confirm that the hypothesis about the influence of society's health status on economic growth is relevant. Except for equation [8] life expectancy affects the explained variable favourably and significantly in statistical terms. It is worth noting, however, that this variable is inserted also as its square only in equation [9]. Bearing in mind that extending life expectancy also increases the economic dependency ratios such modification seems fully justified.

Table 5.2. Estimates of more recent *a la* Barro models of growth

Variable	Equation										
	[1]	[2]	[3]	[4]	[5]	[6]	[7]	[8]	[9]	[10]	[11]
	Source[a]										
	BL[94]	BL[94]	BL[94]	BL[94]	BS[95]	B[97]	SW[97]	Br[98]	Br[98]	Br[98]	B[01]
ODZ for [1]-[8]: ln*ODZ*	0.0801 (5.76)	0.0829 (5.28)	0.0806 (5.80)	0.0903 (6.10)	0.076 (5.07)	0.0418 (3.01)	0.3 (2.8)				
ODZ_2							-0.0026 (2.3)				
WSM	0.0138 (3.29)	0.0133 (3.09)	0.0136 (3.16)	0.0199 (4.15)	0.0164 (2.83)			0.0162 (3.68)	0.0142 (3.30)	0.0128 (2.84)	
WSWM						0.0098 (3.92)					0.0044 (2.44)
WWM			0.0 (0.0)		0.053 (1.77)						
$H_0 \cdot q_0$					-0.209 (2.16)	-0.005 (3.06)					
WSM·FDI									1.0659 (2.77)	1.0659 (2.67)	
FDI								0.6590 (1.41)		- 0.8489 (1.18)	
Δ*WSM*				0.289 (2.39)	0.0066 (1.02)						
SKM		0.0072 (0.62)									
GED					0.205 (1.90)						
WSK	- 0.0092 (1.96)	-0.008 (1.60)	- 0.0061 (1.22)	- 0.0162 (3.00)	- 0.0102 (1.44)						
WWK			-0.021 (0.88)		-0.071 (1.97)						
Δ*WSK*				-0.453 (2.35)	- 0.0128 (1.54)						
SKK		- 0.0119 (0.73)									
Source of data[b]	BL[93]	BL[93]	BL[93]	BL[93]	BL[93]	BL[96]	SH[91]	BL[93]	BL[93]	BL[93]	BL[00]

[a] BL[94] – Barro, Lee [1994], BS[95] – Barro, Sala-i-Martin [1995], B[97] – Barro [1997], SW[97] – Sachs, Warner [1997], B[01] – Barro [2001].

[b] BL[93] – Barro, Lee [1993], BL[96] – Barro, Lee [1996], SH[91] – Summers, Heston [1991], Br – Borensztein et al. [1998], BL[00] – Barro, Lee [2000].

Notes to table 5.2: *ODZ* – life expectancy, *ODZ_2* – square of life expectancy, *WSM* – average male educational attainment, *WWM* - average female educational attainment, *WSWM=WSM+WWM*, *H* – human capital (in equation [5]: an average of 5 indicators: life expectancy and educational attainment separately for secondary and tertiary education, and for males and females; in equation [6]: *H=WSWM*), *FDI* – foreign direct investments, *SKM* – enrolment rates for secondary education and men, *GED* – share of public expenditure on education in GDP, *WSK* – average female educational attainment of secondary education, *WWK* – average female educational attainment of tertiary education, *SKK* – female enrolment rates for secondary education.

Equations: [1]-[5] - 1965-1985, cross-sectional data of 10-year averages, N=180; [6] - 1965-1990, panel data of 10-year averages in the years 1965-1985, and for a 5-year average from the years 1985-1990, N=251, [7] - 1965-1990, cross-sectional data, N=83, [8]-[10] - 1970-1989, panel data for 10-year averages, N=69, [11] - 1965-1995, panel data for 10-year averages, N=246

Estimation method: [1]-[6] - OLS with fixed effect, [7] - OLS, [8]-[10] – seemingly unrelated regression (see e.g. Green [1993], pp. 487-499), [11] - 3OLS (fixed effect)

Other variables in the regressions: [1]-[11] – logarithm of GDP *per capita* in the starting observation of the samples (-), [1]-[6], [8]-[11] – non-investment public expenditure (-), [1]-[6] – number of revolutions and coups (-), relation of the black market exchange rate to the official exchange rate (-), [1]-[5], [8] – share of investment outlays in GDP ((+) except for equation [5]), [5]-[6], [11] – changes in terms of trade (+); [5] – increment in educational attainment of tertiary education for males (no) and females (o) , [6], [11] – logarithm of fertility (-), dummies for Africa (no), sub-Saharan Africa (no), Latin-America (no), Far-East (no); [6] – democracy index (+) and its square (-), [7],[11] – economic openness index (+), product of the economic openness index and logarithm of GDP *per capita* in the first observation of the sample (-), [7] – rate of growth of labour force (+), budget balance (+), index of the quality of public institutions (+), 0-1 – tropic countries (-), 0-1 – countries without access to the sea, share of exports of raw materials in GDP (-), [8]-[11] – logarithm of GDP *per capita* in the starting observation of the sample (+) and its square (-), index of law compliance (+).

Due to multicollinearity between human capital proxies that are separate for males and females, the estimates at female human capital exhibit wrong signs. In equation [4] economic growth is affected by the changes in the level of human capital and the level itself. Equation [5] reports statistical significance of the influence of public expenditure on education. It is worth emphasizing that the significance of the variables under consideration was confirmed in the models with a large number of other explanatory variables, which care that are enumerated in the notes to table 5.2. This fact attests implicitly to the high robustness of variables approximating human capital to changes in the specifications of the growth equations.

Stock variables measuring human capital (educational attainment) behave much better in the analyzed models than stream variables (enrolment rates) do. The statistical significance of the latter has not been even proved.

The higher the level of human capital, the more strongly foreign direct investments influence economic growth (see equations [9] and [8]). Moreover, if the threshold level of human capital is not reached, FDI may even impede growth, because they are directed towards trade and stimulate consumer imports.

5.4.3. The Mankiw-Romer-Weil convergence models and their modifications

Numerous empirical applications rest on the Mankiw-Romer-Weil convergence model as described by formulae (5.31)-(5.32), or on its modifications that assume identical rates of growth of exogenous technological progress for all countries in the sample. The results provided by such models largely depend on the way the source data have been transformed and on estimation methods. For cross-sectional data based on multi-year averages or level specifications (see (5.13)) without the fixed effect the results usually reveal significance of human capital for economic growth. However, the significance is usually lost when the fixed effect methods or differenced data (see 5.32) are used, or outliers are eliminated from the sample; sometimes even the sign of the estimates is changed (from positive to negative).

Table 5.3 reports the results that M-R-W obtained in their original research [1992], as well as outcomes of some other investigations based on the approach recommended by M-R-W, using cross-sectional or multi-year averages. Conclusions offered by the contents of table 5.3 are not univocal. On the one hand, equations [1]-[9] indicate that human capital significantly affects economic growth, but on the other, equations [10]-[14] do not confirm such an impact. In equation [1]-[10] elasticities of production with respect to human capital range from 0.2 to 0.26, whereas with respect to fixed capital they vary from 0.31 to 0.5, which roughly corresponds to the shares of these factors in GDP.

Introduction of an additional production factor in the form of R&D capital reinforces the impact of human capital on the rate of production growth (see variants [5] and [4]). Further extension of the list of the explanatory variables in the modified M-R-W model to government expenditure, unemployment rate, labour force activity rate, as well as group dummies does not blur the significance of human capital for explaining economic growth (see [9]).

Table 5.3. Estimates of the Mankiw-Romer-Weil types of models

Parameter	Equation													
	[1]	[2]	[3]	[4]	[5]	[6]	[7]	[8]	[9]	[10]	[11]	[12]	[13]	[14]
	Source													
	MRW [1992]	MRW [1992]	MRW [1992]	Licht. [1992]	Licht. [1992]	N-V [1996]	V-C [1997]	V-C [1997]	DF [1998]	Temple [1998a]	Temple [1998a]	Temple [1998a]	H-M [1998]	Temple [1998b]
β	0.0186 (9.79)	0.0206 (10.30)	0.0142 (7.45)	0.017 (17.99)	0.021 (4.20)	0.029	0.024	0.033	0.034 (5.25)	0.014	0.015	0.14	0.047	
α_K	0.44 (6.29)	0.38 (2.92)	0.48 (6.86)	0.474 (10.09)	0.354 (4.12)	0.35			0.301 (5.07)				0.468 (5.57)	
α_H	0.23 (3.83)	0.23 (2.09)	0.23 (4.60)	0.236 (4.21)	0.259 (3.65)	0.148			0.204 (3.74)				-0.121 (1.53)	
α_{RD}					0.066 (2.54)	0.084			0.060 (2.22)					
$\ln S_K$	0.506 (5.33)	0.396 (2.61)	0.396 (2.61)			0.413 (2.65)	0.550 (2.90)	0.491 (3.61)		0.66 (5.50)	0.13 (0.65)	0.59 (6.56)		
$\ln S_H$	0.266 (3.33)	0.236 (1.67)	0.236 (1.67)			0.175 (1.55)	0.621 (3.37)	0.558 (3.60)		0.0 (0.0)	0.13 (0.76)	-0.01 (0.17)		0.0 (0.08)
$\ln S_{RD}$						0.098 (1.78)		0.099 (2.25)						
Human capital measure	Gross enrolment rates for secondary education	see [1]	see [1]	see [1]	see [1]	see [1]	Weighted average of gross enrolment rates	see [4]	Enrolment rates for secondary and tertiary education	see [1]	see [1]	see [1]	see [1]	see [1]
Data	S+H[91]	S+H[91]	S+H[91]	S+H[91]	S+H[91]	S+H[91]	S+H[91]	S+H[91]	S+H[91]	S+H[91]	S+H[91]	S+H[91]	S+H[91]	S+H[91]
Estimation method	Restricted OLS	Restricted OLS	Restricted OLS	OLS	OLS	Restricted OLS	OLS	OLS	OLS	LTS + OLS	LTS + OLS	LTS + OLS	OLS – first dif.	OLS
Sample	1960-85 cross-sectional data, non-OECD countries N=75	1960-85 cross-sectional data, OECD countries N=22	1960-85 cross-sectional data, all countries N=98	1960-85 cross-sectional data, N=53	1960-85 cross-sectional data, N=53	1960-85 cross-sectional data OECD countries, N=22	1960-85 cross-sectional data OECD countries, N=22	1960-85 cross-sectional data OECD countries, N=22	1965-95 panel data (5-year averages), OECD countries, N=114(19x6)	1960-85 cross-sectional data, N=69	1960-85 cross-sectional data, OECD countries N=21	1960-85 cross-sectional data, N=92	1960-85 cross-sectional data, N=98	1960-85 cross-sectional data, developing countries, N=58

Notes to table 5.3:

1) Source: MNW [1992] – Mankiw, Romer, Weil; Licht. [1992] – Lichtenberg; N-V [1996] – Nonneman, Vanhoudt; V-C [1996] – Vasudeva, Chien; DF [1997] – de la Fuente; H-M [1998] – Hamilton, Monteadudo; S-H [1998] – Summers, Heston [1991];
2) Other variables: equation [10] – share of public expenditure in GDP (-), changes in unemployment rate (-), economic activity ratio (+), 0-1 for technologically backward countries (-); equations [11], [12], [13] – group 0-1 for African, Latin-American, Far-Eastern and industrial countries; equation [14] – investment outlays on machinery and equipment (+) and on buildings and constructions (+), and 0-1 for African, Latin-American and Far-Eastern countries.

A combined measure of human capital with enrolment rates for all educational stages enables further improvement of the obtained results (see [2] and [7], as well as [6] and [8]).

In all variants reported in table 5.3, statistical significance of the convergence effect associated with parameter β, can be observed. Assuming that all countries have free access to the latest technology, the dynamics of economic growth is higher in countries with lower initial levels of GDP *per capita* by 1.5-3 percentage points than among the technology leaders. Moreover, differences in the estimates of parameter β seem to suggest club convergence (see [2] and [3]).

The results obtained for the Mankiw-Romer-Weil type of models are very sensitive to the presence of outliers, as Temple's research has proved [1998a]. Using the same database as M-R-W did, Temple first identifies outliers by means of the least trimmed squares method (Rousseeuw [1984]), and then runs an estimation procedure taking advantage of the reduced sample. The outcomes show insignificance of human capital (approximated by the enrolment rates for secondary education) for economic growth (see [10], [11], [12]). Furthermore, disaggregation of the investment outlays on machinery and equipment, and construction and buildings also indicates insignificance of human capital (see [14]).

Negative conclusions also follow from the investigation conducted by Hamilton and Monteagudo [1998], in which the researchers used the same database as the M-R-W did, and estimated their model on 10-year average differences between two time spans: 1975-1985 and 1960-1970. The sign at human capital changes into negative, the variable itself being statistically significant (see [13]). According to Jones [1996] this is caused by the exponent – not exponential – introduction of human capital into the production function (see formulae (5.22)-(5.25)).

Starting with a theoretical model accentuating innovation and technological diffusion, Jones constructs an empirical model of growth, in which the logarithm of GDP *per capita* is explained by the logarithm of the share of investment outlays in GDP, and by human capital level. Additionally, Jones analyzes variants in which the level of human capital is replaced with its logarithm. The computations are run for two years: 1960 and 1990, as well as for the differenced variables between these two years. Outcomes reported in table 5.4 seem to confirm Jones' observations. If human capital is introduced into the model in the form of logarithms, then its influence remains positive and significant even in the regression in differences (see [3] and [6]).

The above results, based on the M-R-W model and its modifications, were obtained under a hardly realistic assumption about the non-existence of differences in total factor productivity across the sample countries, the technical manifestation of which was constancy of the intercept by object and application of mainly cross-sectional data.

The results presented in table 5.5 show that letting the intercept vary – to take account of the technological differences between particular countries – leads to estimates that reveal not only the insignificance of human capital for growth, but in many cases they also result in wrong signs. Such results were obtained despite various fixed effect techniques, estimation methods or human capital measures. However, the authors' conclusions do not reject a role of human capital

Table 5.4 Estimates by Jones [1996]

Parameter	Equation					
	[1]	[2]	[3]	[4]	[5]	[6]
$\ln S_K$	0.425 (2.85)	0.437 (2.60)	0.394 (4.15)	0.506 (3.95)	0.377 (2.73)	0.353 (3.72)
H	1.032 (5.61)	0.500 (3.65)	-0.050 (0.39)			
H				0.191 (6.16)	0.189 (6.10)	0.159 (2.48)
α_K	0.298 (4.08)	0.304 (3.75)	0.282 (5.76)	0.336 (6.00)	0.274 (3.75)	0.261 (5.02)
α_H	0.724	0.348	-0.036			
θ				0.127	0.137	0.118
Specification	Levels for 1960	Levels for 1990	Differences	Levels for 1960	Levels for 1990	Differences
Human capital measure	Average educational attainment	see [1]	see [1]	see [1]	see [1]	see [1]
Data source	Summers-Heston [1991] Barro-Lee [1993]	see [1]	see [1]	see [1]	see [1]	see [1]
Sample	N=78	N=78	N=78	N=78	N=78	N=78
Estimation method	OLS	OLS	OLS	OLS	OLS	OLS

in economic growth; they claim instead that such unexpected results may follow from imperfect proxies (streams) of human capital or/and from the fact that the variable may be affecting economic growth indirectly, via TFP.

Table 5.5. Estimates of the Mankiw-Romer-Weil type of models and fixed effect techniques

Parameter	Equation							
	[1]	[2]	[3]	[4]	[5]	[6]	[7]	[8]
	Source							
	Knight et al. [1993]	Knight et al. [1993]	Islam [1995]	Islam [1995]	Islam [1995]	Islam [1995]	Caselli Esquivel Lefort [1996]	Caselli Esquivel Lefort [1996]
β			0.007 (2.76)	0.038 (4.03)	0.016 (2.95)	0.091 (5.71)	0.0107 (3.96)	0.0679 (3.30)
α_K			0.801 (15.01)	0.522 (8.14)	0.602 (5.93)	0.207 (1.97)	0.496 (6.44)	0.491 (4.31)
α_H			0.0544 (0.53)	-0.199 (1.81)	0.017 (0.10)	-0.045 (0.31)	0.18 (3.33)	-0.259 (2.09)
$\ln S_K$	0.105 (10.16)	0.023 (1.61)						
$\ln S_H$	-0.111 (13.26)	-0.065 (5.09)						
Human capital measure	Gross enrolment rates for secondary education	see [1]	Educational attainment	see [3]	see [3]	see [3]	see [1]	see [1]
Data source	S-H [1991]	see [1]	S-H [1991] B+L [1993]	see [3]	see [3]	see [3]	see [1]	see [1]
Sample	1960-85 panel data, 5-year averages	see [1]	see [1] N=79	see [1] N=49	see [1] N=22 (OECD)	see [1] N=22 (OECD)	see [1]	see [1]
Estimation method	fixed effect	fixed effect	OLS	fixed effect	OLS	fixed effect	OLS	IV

To confirm these hypotheses, Knight, Loayza, Villonueva [1993] formulated a variant, in which human capital influences growth positively and significantly, if this variable is inserted into the regression as an average level of human capital for the whole sample[6]. Islam [1995] mentioned in turn a high correlation between the measure of human capital he had used in his analysis and the estimates of total factor productivity.

Table 5.6. Estimates obtained using production functions

Parameter	Equation							
	[1]	[2]	[3]	[4]	[5]	[6]	[7]	[8]
	Source							
	Kyriacou [1991]	Benhabib Spiegel [1994]	Pritchett [1997]	Pritchett [1997]	Pritchett [1997]	Temple [1999b]	Temple [2001]	Temple [2001]
α_K	0.449 (5.05)	0.457 (5.38)	0.524 (12.8)	0.501 (15.4)	0.460 (10.18)	0.553 (13.16)	0.432 (5.08)	0.490 (8.18)
α_L	0.261 (0.90)	0.209 (1.01)				0.241 (2.15)	0.266 (1.38)	
α_H	-0.152 (1.68)	0.063 (0.80)	-0.049 (1.07)	-0.104 (2.07)	-0.120 (1.42)	0.165 (4.00)		
θ							0.015 (0.52)	0.080 (2.56)
Human capital measure	Educational attainment	see [1]	see [1]	see [1]	see [1]	see [1]	see [1]	see [1]
Data source	Kyriacou [1991]	Kyriacou [1991]	Barro Lee [1993]	Nehru Swanson Dubey [1995]	Barro Lee [1993]	Kyriacou [1991]	Kyriacou [1991]	Barro Lee [1993]
Sample	Cross-sectional data 1970-85 N=87	Cross-sectional data 1965-85 N=78	Cross-sectional data 1960-87 N=91	Cross-sectional data 1960-87 N=79	Cross-sectional data 1960-87 N=70	Cross-sectional data 1965-85 N=64	Cross-sectional data 1960-87 N=78	Cross-sectional data 1960-87 N=91
Estimation method	OLS	OLS	OLS	OLS	IV	LTS + OLS	LTS + OLS	LTS + OLS

Notes:

- Omission of α_L means that its estimate was obtained by imposing a restriction of summation of parameters α_K and α_L to unity (constant returns to scale)

[6] This eliminates its variability in time.

5.4.4. Human capital in the production functions

Another group of empirical models of growth is derived from the multi-factor production functions, usually from thc Cobb-Douglas function (see formulae (5.20)-(5.25)), or its modifications. In these models, human capital is regarded as an additional production factor, beside physical capital and labour, that influences economic growth via changes in its stock.

In most investigations of this type, the impact of human capital on GDP *per capita* is found insignificant, and sometimes even negative (see table 5.6). Furthermore, the results are robust to changes in human capital measures (see Benhabib, Spiegel [1994]) and to data sources (see Pritchett [1997]). On the other hand, however, as Temple shows [1999], [2001], acceptable results can be obtained by eliminating the outliers.

Temple [1999], who used the same data as Benhabib, Spiegel [1994] did and reduced the sample from 78 to 64 observations, presents estimates that are consistent with the theoretical postulates (see [2] and [7]). Outliers mainly refer to the OPEC and developing countries that are involved in wars, which – at least partly – explains the lack of relationship between human capital and economic growth. Temple [2001] also experiments with an alternative production function, to which he introduces human capital in the exponential form ([7]-[8]). The results show that, firstly, human capital is significant even if the whole sample is kept (see [3] and [8]), and secondly, that the databases used by Barro, Lee [1993] are more credible than Kyriacou's [1991]. In the first case human capital proved statistically significant (see [7] and [8]).

The empirical results presented in table 5.6 that were obtained using the Cobb-Douglas multi-factor production function do not seem to justify treating human capital as a separate production factor.

Based on his research results, Pritchett [1997] even suggests revising the theoretical concepts of human capital, whereas Kyriacou [1991] and Benhabib, Spiegel [1994] react to their unexpected outcomes by changing the specifications of their equations (see table 5.7). Therefore, Kyriacou [1991] claims that human capital should enter the growth equation as the levels or logarithms, but not as first differences (see [1] and [2]). This hypothesis proves right: the variable is insignificant when inserted as first differences, but significant when introduced as levels or logarithms (see [1] and [2]). It is worth noting that equation [2] indicates the existence of some maximum of human capital's influence on growth, as evidenced by the negative sign at the square of the human capital variable. If this variant were credible, then there would exist some limits to the economic effectiveness of human capital accumulation in an economy.

Table 5.7. Estimates obtained using the modified production functions

Parameter or variable	Equation [1]	[2]	[3]	[4]	[5]	[6]	[7]
	Source						
	Kyriacou [1991]	Kyriacou [1991]	Benhabib Spiegel [1994]	Benhabib Spiegel [1994]	Benhabib Spiegel [1994]	Benhabib Spiegel [1994]	Benhabib Spiegel [1994]
α_K	0.435 (4.88)	0.417 (6.24)	0.479 (5.10)	0.4723 (6.59)	0.5005 (6.49)	0.5076 (5.38)	0.5233 (3.66)
α_L	0.176 (0.58)	0.387 (1.49)	0.391 (2.01)	0.1880 (1.15)	0.2045 (1.31)	0.1720 (0.74)	0.2901 (0.57)
α_H	0.018 (0.12)	0.036 (0.34)					
q_0			-0.235 (5.11)				
H		0.010 (3.25)		-0.0014 (0.09)	0.0021 (0.14)	0.0439 (1.96)	-0.0736 (1.260)
H^2		-0.001 (3.07)					
H	0.0068 (1.79)		0.167 (3.09)				
$H \cdot (QB/Q_i)$				0.0011 (5.50)	0.0007 (2.33)	0.0003 (0.33)	0.0012 (4.00)
QB/Q_i					0.0014 (1.40)		
Human capital measure	Educational attainment	see [1]	See [1]	see [1]	see [1]	see [1]	see [1]
Data source	Kyriacou [1991]	see [1]	See [1]	see [1]	see [1]	see [1]	see [1]
Sample	Cross-sectional data 1970-85 N=87	see [1]	Cross-sectional data 1965-85 N=78	see [3]	see [3]	Cross-sectional data 1965-85 N=26 (countries of low GDP per capita)	Cross-sectional data 1965-85 N=26 (countries of high GDP per capita)
Estimation method	OLS	OLS	OLS	OLS	OLS	OLS	OLS

Notes: Equation [2] contains 0-1 variables for oil exporting countries and Latin-American countries, as well as political stability index; equation [3] contains 0-1 variables for African and Lain-American countries.

In reply to the unsuccessful attempt at finding empirical evidence of the influence of human capital on the rate of growth (see table 5.6, equation [2]), Benhabib and Spiegel augmented the list of explanatory variables to account for effects related to both technological diffusion and convergence. To this end, they introduced a logarithm of human capital and a logarithm of GDP *per capita* in the starting year of the sample, i.e. 1965, to the growth equation. This modification offers acceptable results (see [3] in table 5.7). In the other variants the authors tried to analyze the possible interactions between human capital and technological gap, forming a hypothesis that only countries with adequately large human capital stock are able to benefit from the effect of convergence. Results in variants [4] and [5] confirm this hypothesis: estimates at variable H are insignificant, whereas at the interaction variable – significant. Interestingly, it follows from equation [5] that for the whole world (a whole sample) neither a high level of human capital nor large technological gap (QB/Q_i - where QB denotes GDP *per capita* in the USA, whereas Q_i - GDP *per capita* in the *i*-th country) can guarantee high rates of growth. This somewhat unexpected result changes qualitatively when the whole sample is split up into sub-samples by GDP *per capita*. Equations [6] and [7] show then that, as expected, effects of innovation prevail in the developed countries (significance of variable H), whereas those connected with technological diffusion and absorption of foreign technologies predominate in the developing countries (significance of variable QB/Q_i).

5.4.5. Empirical models with human capital quality and dynamic production functions

An extensive investigation based on standardized international tests as human capital proxies but using a reduced form of the relationship (excluding labour force, physical capital or share of investment in GDP) between human capital and economic growth, was performed by Hanushek and Kimko [2000] (see also section 5.2). The authors, taking advantage of the aforementioned information, constructed two alternative measures of human capital quality, and introduced them to a growth model. They carried out their analysis for 31 countries for which appropriate data were available, and for 80 countries for which relevant data had to be generated (see table 5.8).

As the results in table 5.8 show, the qualitative measures clearly outperform the quantitative measures (educational attainment proves insignificant, whereas both measures recommended by Hanushek and Kimko are statistically significant). What is more, their measures are robust to the introduction of other indicators of human capital (equations [7] and [8]). The authors also proved that causation runs from human capital to economic growth, not vice versa. This conclusion is underpinned by the Mincer wage equation explaining average wages of immigrants in the USA. It turns out that immigrants arriving from countries with a higher quality of education earn in excess of persons coming from countries where its quality is lower.

The reported results give rise to the question about determinants of the quality of human capital. Lee and Barro [2001] attempted to answer the question by means of macroeconomic analysis (see table 5.9). They used the following variables as the factors of human capital quality:

1. Educational attainment and GDP *per capita*; these are justified on the theoretical grounds, because of the positive externalities resulting from parents' education and wealth that increase the effectiveness of knowledge assimilation by their children.

Table 5.8. Estimates by Hanushek, Kimko [2000]

Variable	Equation							
	[1]	[2]	[3]	[4]	[5]	[6]	[7]	[8]
GDP *per capita* in the starting period	-0.609 (3.27)	-0.472 (4.92)	-0.460 (4.47)	-0.270 (3.14)	-0.382 (4.72)	-0.370 (4.40)	-0.393 (4.14)	-0.368 (3.87)
Educational attainment	0.548 (2.62)	0.103 (0.82)	0.100 (0.68)	0.085 (0.75)	0.127 (1.43)	0.120 (1.25)	0.070 (0.67)	0.065 (0.56)
Measure of human capital quality *H1*		0.134 (5.83)		0.091 (3.96)	0.108 (5.14)		0.112 (5.60)	
Measure of human capital quality *H2*			0.104 (6.93)			0.094 (5.88)		0.100 (6.67)
Students / teachers Elementary education							0.001 (0.04)	0.006 (0.25)
Students / teachers Secondary education							-0.038 (0.86)	-0.038 (0.84)
Share of educational expenditure in GDP							7.388 (0.46)	3.968 (0.26)
Data source	Hanushek Kimko [2000]	see [1]	see [1]	see [1]	see [1]	see [1]	see [1]	see [1]
Sample	1960-90 average value in sample N=31	see [1] N=21	see [1] N=31	see [1] N=25	see [1] N=78	see [1] N=80	see [1] N=76	see [1] N=78
Estimation method	OLS	OLS	OLS	OLS	OLS	OLS	OLS	OLS

2. Selected indicators of the quality of teaching, whose effectiveness had been proven in some other investigations (see section 5.2): the number of students per teacher, real expenditure on education per student, real salaries of teachers and the average length of a school day.

Table 5.9. Estimates by Lee, Barro [2001]

	Equation			
	[1]	[2]	[3]	[4]
GDP *per capita* in the starting period (logarithm)	3.19 (3.00)	4.16 (4.23)	3.41 (3.20)	3.39 (3.42)
Educational attainment	1.33 (4.93)	1.33 (4.94)	1.35 (4.90)	1.17 (4.67)
Students / teachers (logarithm	-0.15 (2.44)		-0.22 (2.54)	-0.19 (3.20)
Real expenditures on education per student (logarithm)		1.06 (1.46)	-1.34 (1.13)	
Real average salaries of teachers	1.62 (1.81)		2.88 (2.09)	1.92 (2.28)
Average length of the school day (hours)	0.01 (0.46)		0.003 (0.14)	-0.02 (0.89)
Data source	Barro, Lee [2000]	Barro, Lee [2000]	Barro, Lee [2000]	Barro, Lee [2000]
Sample	N=214	N=214	N=214	N=197
Estimation method	OLS	OLS	OLS	OLS+fixed effect

The results in table 5.9 show that – excluding the average length of the school day and the educational expenditure – the other variables exhibit statistically significant impact on the quality of human capita. The negative signs at the estimates of real expenditure on education per student and, simultaneously, the positive and statistically significant influence of teachers' salaries and the average number of students per class indicate that educational expenditure that neither increases the relative numerical strength of the teaching staff nor their motivation to work does not improve the effectiveness of the teaching process.

In another research into the determinants of human capital quality, Dessus, [1999] who used a varying parameters model, made variability of the parameter measuring human capital impact on economic growth conditional on the function of educational expenditure and the quotient of the number of students over the number of teachers. This approach seems quite interesting, as it allows for some cost-based measures in the growth model, treating them neither as direct production factors nor as alternatives to more adequate measures of human capital.

One of the methodologically most advanced investigations into the role of human capital in economic growth seems to be the investigation by Bassanini and Scarpetta [2001]. The authors, starting with a theoretical model of growth of the Mankiw-Romer-Weil type, built an empirical model that allowed them to separate the short and long-run effects connected with both human capital and some other production factors:

$$\begin{aligned}\Delta \ln y_{it} = \alpha_{0i} - \rho_i \ln y_{it-1} + \beta_{1i} \ln s_{it} + \beta_{2i} \ln h_{it} - \beta_{3i} \Delta \ln l_{it} + \beta_{4i} T + \\ + \lambda_{1i} \Delta \ln s_{it} + \lambda_{2i} \Delta \ln h_{it} + \lambda_{3i} \Delta^2 \ln l_{it} + \varepsilon_{it}\end{aligned} \quad (5.33)$$

where:

y_{it} - GDP per working-age person (15-64 years old) in the i-th country, in year t,

s_{it} - share of private investment outlays – excluding residential outlays – in GDP produced by the private sector,

h_{it} - human capital approximated by educational attainment per working-age person (source: de la Fuente, Domenech [2000]),

l_{it} - number of the working-age persons,

T – linear trend,

α_{0i} - intercept (variable in objects),

ρ_i - coefficient of autoregression,

β_{1i}, β_{2i}, β_{3i}, β_{4i} - long-run parameters,

λ_{1i}, λ_{2i}, λ_{3i} - short-run parameters,

ε_{it} - error term.

Unlike all the quoted investigations, model (5.33) allows all the structural parameters to vary by object/country. The research sample covers 21 of the most developed OECD countries in the years 1971-1998 (altogether 540 observations), and the estimation technique is mean group estimators (see Evans [1997]). The results are presented in table 5.10.

Table 5.10. Estimates by Bassanini, Scarpetta [2001]

Equation	Estimate							
	$\hat{\rho}$	$\hat{\beta}_1$	$\hat{\beta}_2$	$\hat{\beta}_3$	$\hat{\beta}_4$	$\hat{\lambda}_1$	$\hat{\lambda}_2$	$\hat{\lambda}_3$
[1]	-0.37	-0.19		-9.73	0.022	0.12		0.75
	(7.4)	(0.51)		(2.07)	(1.83)	(6.00)		(1.42)
[2]	-0.11	-0.19		-63.78		0.14		0.70
	(10.0)	(0.25		(1.21)		(7.00)		(1.67)
[3]	-0.48	0.26	1.16	-10.08	-0.001	0.09	1.57	1.07
	(8.0)	(2.89)	(0.55)	(1.91)	(0.05)	(4.50)	(0.74)	(2.06)
[4]	-0.36	0.09	1.93	-8.44		0.11	-1.76	0.83
	(7.2)	(0.82)	(3.64)	(1.78)		(3.67)	(2.63)	(1.43)

All the estimates have correct signs. The following variables are statistically significant in each variant: lagged endogenous variable – introduced to account for the convergence effect – and the share of investment outlays in GDP. A clear trade-off between the trend and the human capital can be observed: simultaneous introduction of the two makes them both insignificant. This may attest to strong multicollinearity between these two variables, but it also supports the hypothesis that the main vehicle of technological progress is human capital itself. It is also worth noting that in the most promising variant [4] the short-run effects of increasing human capital on economic growth are negative, but in the long-run expanding human capital stimulates growth.

5.4.6. Human capital and total factor productivity

An alternative concept in investigating the pro-growth effects of human capital rests upon the assumption that it influences economic growth not only directly, but also via total factor productivity (TFP). This means that a variable approximating human capital may enter a growth model, but also a model that accounts for TFP variability. Two different approaches can be distinguished here.

Under the first approach, TFP is computed using the growth accounts. Assuming that the production factors are paid for in accordance with their marginal productivities, the elasticity of production with respect to physical capital can be obtained through calibration, presuming that it is equal to the share of profits from the physical capital in GDP. On the other hand, when the returns to scale are not available, which is a standard assumption in the procedure under consideration, the elasticity of production with respect to labour force is obtained as a complement to unity. Thus, total factor productivity is defined as:

$$TFP = \frac{Y}{K^{\alpha} L^{(1-\alpha)}} \tag{5.34}$$

where:
TFP – total factor productivity,
Y – production (GDP),
K – physical capital,
L – labour force,
α - elasticity of production with respect to physical capital.

At the first stage of the other approach, the structural parameters of a two- or three-factor production function (with human capital in the latter case) are estimated. Next, TFP is computed as a residual between production and production factors weighted by the appropriate elasticities estimated in stage 1. The empirical application of this procedure usually leads to a labour productivity function derived from the Cobb-Douglas type production function by dividing its both sides and then taking logarithms. For a three-factor production function TFP is calculated as:

$$\ln TFP = \ln y - \alpha \ln k - \gamma \ln h - (\alpha + \beta + \gamma - 1) \ln L \tag{5.35}$$

where:
y – labour productivity,
k – capital labour ratio,
h – human capital per worker,
α, β, γ - structural parameters of the Cobb-Douglas production function.

Equation (5.35) generates data on TFP, and the variable is then regressed on various other factors, including human capital (see formula (5.30)).

Representative investigations illustrating empirical applications of the first approach are the works by Engelbrecht [1997], [2002], whereas analyses by Miller, Upadhyay [2000], as well as Tokarski, Roszkowska, Gajewski [2005] are connected with the other approach.

Engelbrecht [1997] augmented the empirical model elaborated by Coe, Helpman [1995] to explain TFP variability using effects related to human capital. In this model, TFP is a function of both domestic stock of high-technology capital, *RD*, defined as accumulated investment outlays on R&D, and foreign stock of such capital, *RF*, weighted by the volumes of imports from particular highly developed countries in the total import of the importing country, *m*. Besides, Engelbrecht introduced into the growth equation a variable approximating the technological gap between a given country and the technologically most advanced economy (Q_i / QB, where *QB* denotes GDP *per capita* in the USA, and Q_i - GDP *per capita* in the *i*-th country). The results of the discussed research are shown in table 5.11.

In all the reported variants, the influence of high-tech capital[7], both domestic and imported, on TFP is statistically significant. The impact of human capital accumulation (equation [1]) is also confirmed and thus its interaction with the technological gap (equations [3] and [4]). It proved necessary to introduce the interaction variable, because, as equation [2] shows, the presence of the level of human capital alone does not guarantee its significance.
Continuing his investigations into the role of human capital and technological diffusion in economic growth, Engelbrecht [2002] formed a hypothesis accentuating the significance of economy's openness / competitiveness for the full realization of the effects of human capital and accumulated outlays on R&D. He used the share of imports in GDP, *M*, as a proxy of economic openness. This variable not only constitutes a separate regressor explaining TFP, but also enters the equation as interaction with both human capital and imported high-technology capital.

[7] Let us recall that in the model under investigation physical capital directly affects the growth dynamics (see formula 5.34).

Table 5.11. Outcomes of TFP estimation by Engelbrecht [1997], [2002] (explained variable: $\Delta \ln TFP$)

Parameter	Equation										
	[1]	[2]	[3]	[4]	[5]	[6]	[7]	[8]	[9]	[10]	[11]
	Source										
	Eng. 1997	Eng. 1997	Eng. 1997	Eng. 1997	Eng. 2002	Eng. 2002	Eng. 2002	Eng. 2002	Eng. 2002	Eng. 2002	Eng. 2002
$\Delta \ln RD$	0.072 (5.29)	0.098 (6.83)	0.098 (6.70)	0.105 (7.15)							
$\ln RD$											
$\Delta m \ln RF$	0.198 (3.93)	0.303 (5.75)	0.249 (4.56)	0.249 (4.72)	-1.829 (1.15)	-1.385 (1.24)					
Δh	0.136 (3.89)										
h		-0.007 (0.42)		0.128 (2.71)							
H									-0.005 (0.83)		0.026 (2.00)
ΔH					0.003 (0.21)	-0.053 (0.18)	0.072 (0.27)				
$\Delta(H \cdot M)$								0.186 (2.21)			
$h \cdot \ln \frac{Q_i}{QB}$			0.141 (2.93)	0.107 (2.82)							
$H \cdot \ln \frac{Q_i}{QB}$									0.007 (2.09)	0.005 (2.33)	0.025 (3.125)
$\ln \frac{Q_i}{QB}$			0.127 (3.34)	-0.260 (4.51)							
ΔM					0.300 (2.07)	-5.338 (1.71)	-5.633 (1.76)	-8.972 (2.27)	-6.315 (2.15)	-5.622 (1.99)	-14.589 (4.88)
$\Delta(M \cdot m \ln RF)$						0.460 (1.78)	0.490 (1.86)	0.761 (2.26)	0.547 (2.27)	0.491 (2.17)	1.208 (0.246)
$\Delta(H \cdot m \ln RF)$						0.005 (0.02)	-0.005 (0.23)				
$\Delta[(H \cdot M) \cdot (M \cdot m \ln RF)]$								-0.034 (4.25)			
Human capital measure	Educ. attain.	see [1]	see [1]	see [1]	see [1]	see [1]	see [1]	see [1]	see [1]	see [1]	Educ-cational attain-ment
Data source	Barro Lee [1993]	see [1]	see [1]	see [1]	Barro Lee [1996]	see [5]	see [5]	see [5]	see [5]	see [5]	see [5]
Sample	21 OECD countries 1971-85 panel data N=315	see [1]	see [1]	see [1]	61 countries 1971-90 5-year averages, N=244	see [5]	see [5]	see [5]	see [5]	see [5]	see [5]
Estimation method	GLS + fixed effect	see [1]	see [1]	see [1]	OLS fixed effect	see [5]	see [5]	see [5]	see [5]	see [5]	see [5]

An additional variable in the model is – just like in the previous Engelbrecht's research [1997] – the product of human capital and technological gap, and this variable corresponds to the convergence effect. This time

Engelbrecht decided in all his variants on the levels or first differences of human capital instead of logarithms. Selected variants of his analysis are given in table 5.11.

The attempt to explain total factor productivity by means of technological diffusion, economic openness, and human capital – all treated as separate variables – fell flat (equation [5]). The variable representing interaction between openness, human capital and imported technology (equations [6] and [7]) did not produce acceptable outcomes, either (negative signs at $\Delta m \ln RF$ and ΔH in equation [6]; improper sign at $\Delta(H \cdot m \ln RF)$ in equation [7]; non-significance of most of the variables). The results became satisfactory (equation [8]) only when Engelbrecht took advantage of what he calls *policy-conditioned human capital stock*[8], and used interactions defined in variant [7]. Alternatively, taking account of the effects of conditional convergence by inserting interaction between human capital and technological gap also leads to correct outcomes, without having to resort to modifications of human capital measure (equations [9]-[11]). However, the sign at human capital level is positive only in equation [11], the variable itself being statistically significant.

The work by Miller and Upadhyay [2000] exemplifies an attempt at computing total factor productivity using the econometric techniques, with a consequent effort to explain TFP by, *inter alia*, human capital. The authors first estimated two variants of the Cobb-Douglas production functions: a two-factor function without human capital, and a three-factor one with human capital. The outcomes are as follows:

a) for the two-factor function:

$$\ln \hat{y}_{it} = 0{,}4756 \ln k_{it} - 0{,}0988 \ln L_{it} + \sum_{t=1}^{6} \hat{\theta}_t \cdot U_t \tag{5.36}$$

(18,86) (1,32)

b) for the three-factor function:

$$\ln \hat{y}_{it} = 0{,}4712 \ln k_{it} - 0{,}1383 \ln L_{it} + 0{,}1136 \ln h_{it} + \sum_{t=1}^{6} \hat{\omega}_t \cdot U_t \tag{5.37}$$

(18,64) (1,78) (1,81)

where:

U_t - dummy accounting for intercept changes in time (six 5-year averages),

θ, ω - parameters accounting for intercept variability in time.

The data used in the analysis cover 83 countries and 5-year averages for the years 1960-1989. The estimation was run using the OLS and each variable was measured as a deviation from its average. Consequently, equation (5.37)[9]

[8] This variable is a product of human capital and economic openness, *HM*.

[9] For the case of a two-factor function the procedure is analogous.

made it possible to compute residuals (TFP) for each observation in the sample in the following manner:

$$\ln TFP_{it} = \ln \bar{y}_i - 0{,}4712 \ln \bar{k}_i - 0{,}1136 \ln \bar{h}_i - 0{,}1383 \ln \bar{L}_i + \hat{\omega}_t \qquad (5.38)$$

Having arrived at TFP estimates in two ways, Miller and Upadhyay tried to explain its variance applying the same techniques as in the production function: each variable was measured as a deviation from its average; six dummies were introduced; and the estimation method was OLS. The authors formulated a hypothesis that the influence of human capital on TFP was conditional on the degree of economy's openness/competitiveness, measured by the share of exports in GDP, x. Beside human capital and its interaction with the share of exports in GDP, they used the following variables as regressors: terms of trade, *tot*; deviations from purchasing power parity, *pd*; inflation rate, π, and the variables' standard deviations: σ_x, σ_{tot}, σ_{pd}, σ_π. The estimation results of equations explaining the variance of TFP logarithms are presented in table 5.12.

This research shows that the impact of human capital on total factor productivity is evidently positive only above the threshold value of openness. In equations [1] and [3], with no interaction variable between human capital and openness, human capital proves insignificant[10]. It follows from the reported results that in closed economies without any access to the latest technologies – which implicitly weakens their competitiveness – human capital does not contribute to TFP increase. According to the results presented in table 5.12, the following variables also significantly contribute to the variance of TFP: the rate of inflation and the local currency deviations from its purchasing power parity. This confirms the hypothesis that economic growth is determined not only by real processes, but also by the financial ones.

Table 5.12. Results of TFP estimation by Miller, Upadhyay [2000]. Variants [1] and [2] – TFP derived from the two-factor production function; variants [3] and [4] – TFP derived from the three-factor production function

	[1]	[2]	[3]	[4]
$\ln H$	0.0090 (0.29)	-0.0961 (1.64)	-0.0437 (1.39)	-0.1357 (2.32)
$\ln x$	0.1237 (6.08)	0.0925 (3.70)	0.1207 (5.97)	0.0935 (3.76)
$\ln H \cdot \ln x$		0.0395 (2.12)		0.0346 (1.87)
$\ln tot$	0.0283 (1.38)	0.0286 (1.40)	0.0295 (1.45)	0.0298 (1.46)

[10] The sign at human capital in equation [3] is even negative. One should remember, however, that the explained variable in equations [3] and [4] is computed from the production function, in which human capital is an explicit explanatory variable.

$\ln pd$	-0.0703 (2.03)	-0.0821 (-2.35)	-0.0779 (2.26)	-0.0882 (-2.54)
$\ln(1+\pi)$	-0.1550 (3.78)	-0.1560 (3.82)	-0.1548 (3.79)	-0.1557 (3.83)
$\ln\sigma_x$	-0.0023 (3.14)	-0.0028 (3.65)	-0.0024 (3.20)	-0.0028 (3.63)
$\ln\sigma_{tot}$	0.00003 (0.23)	0.00004 (0.30)	0.00003 (0.21)	0.00004 (0.27)
$\ln\sigma_{pd}$	0.0006 (1.60)	0.0008 (1.94)	0.0006 (1.68)	0.0008 (1.97)
$\ln\sigma_{\pi}$	0.0116 (1.33)	0.0117 (1.35)	0.0115 (1.33)	0.0116 (1.34)
Human capital measure	Educational attainment	See [1]	See [1]	See [1]
Data source	Summers-Heston [1991] Barro-Lee [1993]	See [1]	See [1]	See [1]
Sample	1960-89 panel data, 5-year averages	See [1]	See [1]	See [1]
Estimation method	OLS	See [1]	See [1]	See [1]

In conclusion of the discussion about the Miller and Upadhyay's work [2000] it should be noted that the three-factor production function contains elements of tautology, because the effects of human capital were eliminated, when TFP was being computed. Under this definition of total factor productivity, TFP should not be expected to depend on human capital.

Tokarski, Roszkowska, Grajewski [2005], who used panel data for the period 1996-2001 and took advantage of the already described methodology, made an attempt to explain the diversity of total factor productivity in Poland by referring to the voivodeships[11]. Unlike Miller and Upadhyay, in the regression equation they shifted the intercept by object (voivodeship), not in time. They employed two combined variables as the human capital proxies: one being the numbers of tertiary students and academic teachers, and the other being the urbanization ratio for large cities, the length of roads per 1 square kilometre, and also the distance (in Euclidean space) between the labour force structure in Poland and its 1999 counterpart in EU-15. The authors arrived at significant influence of human capital – finally accepting the first combined variable – upon the variance of total factor productivity.

[11] Polish administrative regions.

Both the investigations – by Miller, Upadhyay, as well as Tokarski, Roszkowska, Grajewski – illustrate the multitude of possible variables that can determine total factor productivity, amongst which human capital plays a crucial role.

5.4.7. Time series empirical models of growth and human capital

Most empirical investigations exploring the role of human capital in economic growth rest on cross-sectional or panel data, and the objects are countries or domestic regions. Applications based on time series data on individual countries seem scarcer (see e.g. Dougherty, Jorgenson [1996], Jorgenson, Ho, Stiroh [2003] or Welfe [2001]).

Jorgenson used the growth accounts to compute the contribution of individual production factors – including human capital as given by formula (5.15) – to the economic growth in the USA. Human capital does not constitute a separate production factor there, but enters the calculations as a product with the working population, which can be interpreted in terms of effective labour force. By the same token effective stock of physical capital is computed. Drawing on Clark's marginal theory of division, parameters of the growth model are derived from the appropriate identities by means of the calibration techniques. The outcomes of the computations made by Jorgenson, Ho and Stiroh [2003], where in addition to the impact of human capital also the contribution of other production factors are presented, can be found in table 5.13.

Table 5.13. Factors of growth in the USA in the years 1959-2000 (rates of growth)

Factors of growth	Time interval			
	1959-2000	1959-1973	1973-1995	1995-2000
Rate of growth of GDP	3.61	4.24	2.99	4.60
Time worked	1.54	1.27	1.55	2.24
Labour productivity per hour worked	2.07	2.97	1.44	2.36
Capital / labour ratio	1.13	1.44	0.88	1.40
Capital / labour ratio in IT sector	0.32	0.16	0.32	0.76
Capital / labour ratio in non-IT sectors	0.82	1.28	0.56	0.64
Labour force quality (human capital)	0.27	0.39	0.23	0.13
Total factor productivity (TFP)	0.67	1.14	0.33	0.84
TFP due to IT sector	0.23	0.10	0.24	0.51
TFP due to non-IT sectors	0.44	1.03	0.08	0.33

Source: Jorgenson, Ho, Stiroh [2003]

Data in table 5.13 indicate that the impact of human capital upon the economic growth of the USA is decreasing in time. This unexpected – at first sight – conclusion is fully justified and can be generalized to other highly-developed countries. Namely, in countries typified by already high levels of human capital associated with educational attainment, further dynamic development of this capital is limited.

In the model proposed by Welfe [2001] human capital does not constitute a separate production factor, either, but generates effective labour force instead. The productivity of fixed assets is a function of the capital-labour ratio per time effectively worked, total factor productivity, and a few dummies neutralizing demand shocks. In line with Coe and Helpman's [1995] proposal, TFP is explained by domestic and imported, accumulated outlays on R&D. All parameters of the production function are estimated in one step. Estimation of the machinery and equipment productivity equation for Poland in the period 1976-1998 provides the following results :

$$\begin{aligned} \Delta \ln WXKM_t = {} & -\underset{(4,7)}{0{,}501}\Delta \ln TH_t + \underset{(4,8)}{1{,}075}\Delta \ln WN_t + \underset{(0,5)}{0{,}076}\Delta \ln BIRK_t + \\ & \underset{(1,4)}{0{,}084} BIRM_t + \sum_{i=1}^{3} \hat{\alpha}_i \cdot U_{it} \end{aligned} \qquad (5.38)$$

where:

WXNM – productivity of machinery and equipment;

WN – standardized ratio of time-worked,

BIRK – cumulated domestic outlays on R&D,

BIRM – cumulated imported outlays on R&D weighted by the shares of imports from Poland's highly developed trading partners,

U – dummy variables.

Human capital in equation (5.39) - computed according to formula (5.14) – only participates in defining the capital-labour ratio per time effectively worked, *TH*. The results show that the intercept in equation (5.39) is insignificant; this could be interpreted as a comprehensive specification of the equation, but the concurrent insignificance of variables *BIRK* and *BIRM* may indicate that introduction of human capital as a separate explanatory variable is necessary.

5.4.8. Comparative analysis of the outcomes of investigations into the role of human capital in economic growth

Estimates derived from the empirical models of growth are hardly a basis for formulating a univocal conclusion about the role human capital plays in economic growth. Depending on the specifications and data applied, the results show either insignificance of this variable (sometimes even its adverse influence) or its crucial influence on the long-term growth. As Fuente claims [2004], this surprising range of outcomes can be largely ascribed to the poor quality of the data.

Fuente estimated structural parameters of the three-factor Cobb-Douglas production function with constant returns to scale, taking advantage of the most popular international databases of human capital. The computations were carried out for a few functional forms: the log-linear, with and without fixed effect, and the differenced logarithms, with and without fixed effect. Additionally, Fuente computed the reliability ratios for all his human capital proxies and stated that the highest values of the ratios were obtained for proxies showing significance of human capital in each variant (see table 5.14). Fuente's experiment seems to confirm the fundamental role of data quality in assessing the function of human capital in economic growth: the higher the reliability ratios, the higher the precision of the estimates and significance of human capital.

Another interesting research illustrating impacts of various estimation techniques on the outcomes of the empirical growth models is the study by Uramek [2006].

The author recommends a synthetic taxonomic measure of human capital and verifies its influence on economic growth using a modified M-R-W model and the following estimation methods: OLS, fixed effect, random effect (Green [1993], pp. 464-480), the Arellano-Bond method [1991], and the Blundell-Bond method of moments [1998]. The author concludes that OLS underestimates the impact of human capital on economic growth and therefore the Blundell-Bond method of moments should be preferred.

Table 5.14. Estimates by de la Fuente [2004]

Functional form	Data source								Average
	Kyriacou [1991]	Nehru-Swanson-Dubey [1995]	Barro-Lee [1993]	Barro-Lee [1996]	Barro-Lee [2000]	Cohen-Soto [2001]	De la Fuente-Domenech [2000]	De la Fuente-Domenech [2001]	
Log-linear	0.186 (2.18)	0.078 (2.02)	0.141 (4.49)	0.165 (4.82)	0.238 (6.19)	0.397 (7.98)	0.407 (7.76)	0.378 (6.92)	0.249 (5.30)
Log-linear +fixed effect	0.066 (1.86)	0.068 (0.76)	0.136 (3.30)	0.115 (1.80)	0.203 (3.74)	0.608 (4.49)	0.627 (3.99)	0.958 (6.51)	0.348 (3.31)
Differences in logarithms	0.009 (0.15)	0.079 (0.70)	0.089 (2.52)	0.083 (1.47)	0.079 (1.28)	0.525 (2.57)	0.520 (2.17)	0.744 (3.10)	0.266 (1.75)
Differences in logarithms +convergence effect	0.014 (0.29)	-0.206 (1.61)	0.056 (1.80)	-0.007 (0.11)	-0.019 (0.31)	0.573 (3.52)	0.587 (3.47)	0.540 (2.89)	0.192 (1.24)
Average	0.069 (1.12)	0.005 (0.47)	0.106 (3.03)	0.089 (2.00)	0.125 (2.73)	0.526 (4.64)	0.535 (4.35)	0.655 (4.86)	x

In the light of the two investigations, the favourable contribution of human capital to growth should be acknowledged as indisputable. However, the answer to the question about specific mechanisms within this relationship and its strength is not simple. If we assume that the most reliable results are provided by the latest investigations benefiting from data with high reliability ratios (Fuente [2004]), then the interval of variability of the estimates of growth elasticities with respect to human capital becomes considerably narrower. Average values of these elasticities presented in the last four columns of table 5.14 (Barro-Lee [2000], Cohen-Soto [2001], Fuente-Domenech [2000], [2001]) can be regarded as the most representative. The interval containing growth elasticity with respect to human capital ranges from 0.125 to 0.655 (averages by different functional form from the four reported researches), and from 0.355 to 0.599 (averages by the same functional form from the four reported investigations). Because the second interval fits within the first one, it can be skipped in the following analyses. If the obtained estimates were to be treated as a benchmark to investigations into the role of human capital in the Polish economy, then the interval should be broadened by 1-2 standard deviations. For 1 standard deviation the interval becomes 0.08-0.79.

The outcomes of the various studies presented in chapter 5 make it possible to formulate several generalizations. Firstly, a simultaneous use of a few alternative measures of human capital in empirical growth models usually renders some of them insignificant.

Secondly, if one fails to prove a direct impact of human capital upon growth, a conditional treatment of this variable usually helps corroborate such an impact. In this context, conditionality is meant as the introduction of interactions between human capital and some other variables in order to verify various economic hypotheses (convergence, technological gap, openness).

Thirdly, the proposed empirical models of growth leave out particular features of human capital proxies, e.g. in the M-R-W type of models proxies such as the enrolment rates, educational attainment, wage relations or shares of outlays on education in GDP can be found.

Fourthly, the better the measure of human capital (stock instead of a stream, tests' results instead of educational attainment), the more strongly this variable affects economic growth.

Fifthly, by taking account of sample homogeneity we can improve the precision of the estimates and identify the various mechanisms within the influence of human capital on growth.

An important obstacle to direct comparisons between outcomes of various investigations building on the international cross-sectional or panel data and those using time series data on a specific economy (e.g. Poland) is that they employ different definitions of the explained variable. International research uses rates of growth *per capita*, whereas the time series analyses use variance of GDP per employee as the regressand.

References

Aghion Ph., P. Howitt, 1999, *Endogenous Growth Theory*, MIT Press, London.
Angrist J.D., Lavy V., 2001, *Does teacher training affect pupil learning? Evidence from matched comparisons in Jerusalem public schools*, "Journal of Labour Economics", vol. 19/2, p. 343-369.
Arellano M., Bond S., 1991, *Some Tests of Specification for Panel Data: Monte Carlo Evidence and an Application to Employment Equation*, "Review of Economic Studies", vol. 58, p. 277-297.
Asteriou D., Agiomirgianakis G.M., 2001, *Human capital ad economic growth. Time series evidence from Greece*, "Journal of Policy Modeling", vol. 23, p. 481-489.
Azariadis C., Drazen A., 1990, *Threshold Externalities in Economic Development*, "Quarterly Journal of Economics", vol. 105/2, p. 501-526.
Barro R. J., 1991, *Economic Growth in a Cross Section of Countries*, "Quarterly Journal of Economics", vol. 106/2, p. 407-443.
Barro R. J., 1997, *Determinants of Economic Growth: A Cross-Country Empirical Study*, Cambridge, MA: MIT Press.
Barro R. J., 2001, *Human Capital and Growth*, "American Economic Review, Papers and Proceedings", vol. 91/2, p. 12-17.
Barro R.J., Lee J.W., 1993, *International Comparisons of Educational Attainment*, "Journal of Monetary Economics" vol. 32, p. 363-94.
Barro R.J., Lee, J.W., 1994, *Sources of Economic Growth*, "Carnegie-Rochester Conference Series on Public Policy", vol., p. 1-46.
Barro R.J., Lee J.W., 1996, *International Measures of Schooling Years and Schooling Quality*, "American Economic Review, Papers and Proceedings", vol. 86/2, p. 218-223.
Barro R.J., Lee J.W., 2000, *International data on educational attainment. Updates and implications*, "NBER Working Paper" , no. 7911.
Barro R. J., Sala-i-Martin X, 1995, *Economic Growth*, New York: McGraw-Hill.
Bassanini A., Scarpetta S., 2001, *Does Human Capital Matter for Growth in OECD Countries?: Evidence from Pooled Mean-Group Estimate*, "OECD Economics Department Working Papers", no 282, OECD Publishing.
Baumol W., 1986, *Productivity growth convergence and welfare: what the long-run data show*, "American Economic Review", vol. 76, p. 1072-85.
Baumol W., Blackman B., Wolf E., 1989, *Productivity and American Leadership: the Long View*, MIT Press.

Benhabib J., Spiegel M. M., 1994, *The Role of Human Capital in Economic Development: Evidence from Aggregate Cross-Country and Regional U.S. Data*, "Journal of Monetary Economics", vol. 34/2, p. 143-173.
Bils M., Klenow K.J., 2000, *Does Schooling Cause Growth*, "American Economic Review", vol. 90/5, p. 1160-1183.
Bjoornskov Ch., 2006, *The multiple facets of social capital*, "European Journal of Political Economy", vol. 22, p. 22-40.
Blundell R., Bond S., 1998, *Initial Conditions and Moment Restrictions in Dynamic Panel Data Model*, "Econometric Review", vol. 19/3, p. 321-340.
Borensztein E., Gregorio J., Lee J., 1998, *How does foreign direct investment affect economic growth*, "Journal of International Economics", vol. 45, p. 115-135.
Borghans L., Green F., Mayhew K., 2001, *Skills Measurement and Economic Analysis: An Introduction*,. "Oxford Economic Papers", vol. 53, p. 375-384.
Bucci A., 2003, *R&D, imperfect competition and growth with human capital accumulation, Departamental*, "Working Papers", no. 11, Department of Economics, University of Milan.
Canton E, Minne B., Nieuwenhuis A., Smid B., Steeg M., van der, 2005, *Human capital, R&D, and competition in macroeconomic analysis*, "CPB Document", no. 91.
Caselli F., Esquivel G., Lefort F., 1996, *Reopening the convergence debate: a new look at cross-country empirics*, "Journal of Economic Growth", p. 363-89.
Coe D., Helpman E., 1995, *International R&D Spillovers*, "European Economic Review", vol. 39, p. 859-87.
Cohen D., Soto M., 2001, *Growth and human capital: good data, good results*, "CEPR Discussion Papers", no. 3025.
Dagum C., Slottje D. J., 2000, *A New Method to Estimate the Level and Distribution of Household Human Capital with Application*, "Structural Change and Economic Dynamics", vol. 11, p. 67-94.
Denison E. F., 1967, *Why Growth Rates Differ: Postwar Experience in None Western Countries*, The Brookings Institution, Washington.
Dessus S., 1999, *Human capital and growth: the recovered role of educational systems*, "World Bank Policy Research Working Paper", no. 2632.
Dougherty Ch., Jorgenson D., 1996, *International Comparisons of the Sources of Economic Growth*, "The American Economic Review", vol. 80/2, p. 25-29.
Eisner R., 1985, *The total incomes system of accounts*, "Survey of Current Business", vol. 65/1, p. 24-48
Engelbrecht H.J., 1997, *International R&D spillovers, human capital and productivity in OECD economies: An empirical investigation*, European Economic Review, vol. 41, p. 1479-88.
Engelbrecht H.J., 2002, *Human capital and international knowledge spillovers in TFP growth of a sample of developing countries: a exploration of alternative approaches*, "Applied Economics", vol. 34, p. 831-41.

Evans P., 1997, *How fast do economies converge*, "The Review of Economics and Statistics", vol. 79/2, p. 219-225.

Florczak W., 2007, *Modelownie sektora nauki i edukacji – makroekonomiczne ujęcie systemowe* (*Modeling science and education sectors – a system approach*), in: E. Okoń-Horodyńska, S. Pangsa-Kania (red.), *Strategia Lizbońska a możliwości budowania gospodarki opartej na wiedzy* (*Lisbon Strategy and knowledge-based economy*), PWE, Warszawa, p. 16-32.

Fuente, de la, A., 1998, *Convergence equations and income dynamics: The sources of OECD Convergence*, "CEPR Discussion Paper", no. 1794.

Fuente, de la, A., 2004, *Human Capital and Growth: Some Results for the OECD*, in: *Current Issues of Economic Growth,* Proceedings of OECD Workshops, Oesterreichische Nationalbank, Vienna.

Fuente, de la, A., Domenech R., 2000, *Human Capital in Growth Regressions: How Much Difference Does Data Quality Make?*, "CEPR Discussion Paper", 2466, Centre for Economic Policy Research, London.

Fuente, de la, A., Domenech R., 2001, *Schooling data, technological diffusion and the neoclassical model*, "American Economic Review", vol. 91(2), Papers and Proceedings, p. 323-27.

Gemmell N., 1996, *Evaluating the Impacts of Human Capital Stocks and Accumulation on Economic Growth: Some New Evidence*, "Oxford Bulletin of Economics and Statistics", vol. 58/1, p. 9-28.

Green W., 1993, *Econometric Analysis*, Macmillan Publishing Company, New York.

Gundlach E., 1995, *The Role of Human Capital in Economic Growth: New Results and Alternative Interpretations*, "Weltwirtschaftliches Archiv", vol. 131/2, p. 383-402.

Hall R., Jones I., 1999, *Why do some countries produce so much more output per worker than others?*, "The Quarterly Journal of Economics", vol. 144/1, p. 83-116.

Hamilton, J., Monteagudo J., 1998, *The augmented Solow model and the productivity slowdown*, "Journal of Monetary Economics", vol. 42, p. 495-509.

Hanushek E. A., Kimko D. D., 2000, *Schooling, Labour Force Quality, and the Growth of Nations*, "American Economic Review", vol. 90/5, p. 1184-1208.

Hers J., 1998, *Human Capital and Economic Growth*, "CPB Report", vol. 2, p. 36-41.

Islam N., 1995, *Growth Empirics: A Panel Data Approach*, "Quarterly Journal of Economics", vol. 119/4, p. 1127-1170.

Jones C., 1996, *Human capital, ideas and economic convergence*, Stanford University.

Jorgenson D. W., 1995, *Productivity. Volume 1: Postwar U.S. Economic Growth. Volume 2: International Comparisons of Economic Growth*, MIT Press, Cambridge.

Jorgenson D. W., Fraumeni B.M., 1989, *The accumulation of human and non-human capital, 1948-1984*, in: Lipsey, R.E., Tice H.S. (eds.), *The Measurement of Savings, Investment and Wealth*, The University of Chicago Press, Chicago.
Jorgenson D.W., Ho M.S., Stiroh K.J., 2003, *Lessons from the US growth resurgence*, "Journal of Policy Modeling", p. 453-470.
Jovanovic B., Rob R., 1999, *Solow vs. Solow*, New York University, mimeo.
Kendrick J., 1976, *The Formation and Stocks of Total Capital*, Columbia University Press for NBER, New York.
Klenow P.J., Rodrigues A., 1997, *The Neoclassical Revival in Growth Economics: Has it Gone too Far?*, in: Bernanke B.S., Rotemberg (eds.), *NBER Macroeconomics Annal*, MIT Press Cambridge, p. 73-103.
Knight M., Loayza N., Villanueva D., 1993, *Testing the neoclassical theory of economic growth: A panel data approach*, "IMF Staff Papers", vol. 40/3, p. 512-41.
Krueger A. B., 2003, *Economic Considerations of Class Size*, "Economic Journal", vol. 113, p.34-63.
Krueger A. B., Lindahl M., 2001, *Education for Growth: Why and for Whom?*, "Journal of Economic Literature", vol. 39/4, p. 1101-1136.
Kyriacou G. A., 1991, *Level and Growth Effect of Human Capital: A Cross-Country Study of the Convergence Hypothesis*, Economic Research Reports 19-26, C.V. Starr Center for Applied Economics, New York University.
Landau D., 1983, *Government Expenditure and Economic Growth: a Cross-Country Study*, "Southern Economic Journal", January, p. 783-92.
Landau D., 1986, *Government Expenditure and Economic Growth in the Less Developed Countries: an Empirical Study for 1960-80*, "Economic Development and Cultural Change", vol. 35, p. 35-75.
Laroche M., Merette M., Ruggeri G.C., 1999, *On the Concept and Dimensions of Human Capital in a Knowledge-Based Economy Context*, "Canadian Public Policy", vol. XXV, no. 1, p. 87-100.
Lau L. J., Jamison, D. T., Louat F. F., 1991, *Education and Productivity in Developing Countries: An Aggregate Production Function Approach*, "World Bank PRE Working Paper Series", no. 612, Washington.
Lau L., Bhalla S., Louat F., 1991, *Human and physical capital stock in developing countries: construction of data and trends*, "World Development Report", World Bank, New York.
Le T., Gibson J., Oxely L., 2003, *Cost- and Income-Based Measures of Human Capital*, "Journal of Economic Surveys", vol. 17. vol. 3, p. 271-307.
Leamer E., 1985, *Sensitivity Analyses Would Help*, "American Economic Review", vol. 57/3, p. 308-313
Lee J.W., Barro R., 2001, *Schooling quality in a cross-section of countries*, "Economica", vol. 68, p. 465-488.
Leuven E., Lindahl M., Oosterbeek H., Webbink D., 2004, *New Evidence on the Effect of Time in School Achievement*, "Scolar Working Paper", no. 47/04.

Levine R.E., Renelt D., 1992, *A Sensitivity Analysis of Cross-Country Growth Regression*, "American Economic Review", vol. 82/4, p. 942-963.
Lichtenberg F., 1992, *R&D investment and international productivity differences*, "NBER Working Paper", no. 1813.
Lucas R. E., 1988, *On the mechanics of economic development*, "Journal of Monetary Economics", vol. 22, vol. 1, p. 3-42.
Machlup F., 1984, *The Economics of Information and Human Capital*, vol. 3, Princeton, N.J.: Princeton University Press.
Mankiw N.G., Romer D., Weil D.N., 1992, *A Contribution to the Empirics of Growth*, "Quarterly Journal of Economics", vol. 107/2, p. 408-437.
Miller S.M., Upadhyay M.P., 2000, *The effects of openness, trade orientation, and human capital on total factor productivity*, "Journal of Development Economics", vol. 63, p. 399-423.
Milo W., Szafrański G., 2006, *Uwagi na temat pomiaru kapitału ludzkiego* (*Notes on measuring human capital*), in: Zeliaś A. (red.), *Przestrzenno-czasowe modelowanie i prognozowane zjawisk gospodarczych* (*Time-spatial modeling of economic phenomena*), Wydawnictwo Akademii Ekonomicznej w Krakowie, p. 81-98.
Mincer J., 1970, *The Distribution of Labour Incomes: A Survey. With Special Reference to Human Capital Approach*, "Journal of Economic Literature", VIII (1), March, p. 1-26.
Mulligan C.B., Sala-i-Martin X., 1995, *A Labour Income-Based Measure of the Value of Human Capital: An Application to the States of the United States*, "NBER Working Paper", no. 5018.
Nehru V., Swanson E., Dubey A., 1995, *A New Database on Human Capital Stock in Developing and Industrial Countries: Sources, Methodology, and Results*, "Journal of Development Economics", vol. 46/2, p. 379-401.
Nelson R. Phelps E., 1966, *Investment in humans, technological diffusion, and economic growth*, "American Economic Review", vol. 56/1-2, p. 69-75.
Nonneman W., Vabhoudt P., 1996, A further augmentation of the Solow model and the empirics of economic growth for OECD countries, "Quarterly Journal of Economics", vol. 11, p. 943-53.
O'Neill D., 1995, *Education and Income Growth: Implications for Cross-Country Inequality*, "Journal of Political Economy", vol. 103/6, p. 1289-1301.
Prichett L., 1997, *Where has all the education gone?*, The World Bank, Washington DC, "Policy Research Working Paper", no. 1581.
Psacharopoulos G., 1994, *Returns to Investment in Education: A Global Update*, "World Development", vol. 22 (9), p. 1325-43.
Putnam R., 1993, *Making Democracy Wok: Civic Traditions in Modern Italy*, Princeton University Press, Princeton.
Redding S., 1996, *The low-skill, low-quality trap: Strategic complementaries between human capital and R&D*, "Economic Journal", vol. 106, p. 458-470.

Romer P.M., 1990a, *Endogenous technological change*, "Journal of Political Economy", vol. 98, vol. 5, p. 71-102.
Romer P.M., 1990b, *Human capital and growth: theory and evidence*, "Carnegie-Rochester Conference Series on Public Policy", vol. 32, p. 251-286.
Romer P.M., 2000, *Should the Government Subsidize Supply or Demand in the Market for Scientists and Engineers*, "NBER Working Papers", no. 7723.
Rousseeuw P., 1984, *Least median of squares regression*, "Journal of the American Statistical Association", vol. 79, p. 871-80.
Sachs J., Warner A., 1997, *Fundamental sources of long-run growth*, "American Economic Review", Papers and Proceedings, vol. 87/2, p. 184-188.
Sala-i-Martin X., 1997, *I just ran two million regressions*, "American Economic Review", vol. 87/2, p. 178-183.
Scarpetta S., Bassanini A., Pilat D., Schreyer P., 2000, *Economic growth in the OECD area: recent trends at the aggregate sectoral level*, "OECD Economics Department Working Papers", no. 248.
Shultz T. W., 1961, *Investment in Human Capital*, "American Economic Review", vol. 51/5, p. 1-17.
Siwiński W., 2005, *Międzynarodowe zróżnicowanie rozwoju gospodarczego: fakty i teoria* (*International differentiation of economic development: facts and theory*), „Ekonomista", no. 6, p. 723-746.
Solow R.M., 1957, *Technical Change and the Aggregate Production Function*, "Review of Economics and Statistics", vol. 39, p. 312-320.
Summers R., Heston A., 1988, *A New Set of International Comparisons of Real Product and Price Levels: Estimates for 130 Countries*, "Review of Income and Wealth", vol. XXXIV, p. 1-25.
Summers R., Heston A., 1991, *The Penn World Table (Mark 5): An Expended Set of International Comparisons, 1950-1988*, "Quarterly Journal of Economics", vol. 106/2, p. 327-68.
Sztaudynger J., 2003, *Póby ekonometrycznġo określenia wpływu kapitału społecznego na wzrost gospodarczy* (*An attempt to econometrically determine the influence of social capital on economic growth*), „Gospodarka Narodowa", p. 1-17.
Świtalski W., 2000, *The Knowledge-based Economy Paradigm: From Factographic Research to Model Building*, in: Kukliński A., Orłowski W. (eds.), *The Knowledge-based Economy. The Global Challenges of the 21st Century*, vol. 4, pp. 152-170, KBN, Warszawa
Temple J., 1998a, *Robustness tests of the augmented Solow model*, "Journal of Applied Econometrics", vol. 13, p. 361-75.
Temple J., 1998b, *Equipment investment and the Solow model*, "Oxford Economic Papers", vol. 50, p. 39-62.
Temple J., 1999a, *The New Growth Evidence*, "Journal of Economic Literature", vol. 37, p. 112-156.

Temple J., 1999b, *A Positive Effect of Human Capital on Growth*, "Economics Letters", vol. 65/1, p. 131-134.
Temple J., 2001, *Generalizations that aren't? Evidence on education and growth*, "European Economic Review", vol. 45(4-6), p. 905-18.
The Well-being of Nations: The Role of Human Capital and Social Capital, 2001, OECD, Paris.
Tokarski T., Roszkowska S., Gajewski P., 2005, *Regionalne zróżnicowanie łącznej produktywności czynników produkcji w Polsce (Regional differsification of total factor productivity in Poland)*, „Ekonomista", no. 2, p. 215-244.
Uramek K., 2006, *Kapitał ludzki a wzrost gospodarczy. Taksonomiczne wskaźniki kapitału ludzkiego w wybranych krajach (Human capital and economic growth. Taxonomic indicators of human caital in selected countires)*, in: Krajewski S, Kaczorowski P. (red.), *Wzrost gospodarczy, restrukturyzacja i rynek pracy w Polsce. Ujęcie teoretyczne i empiryczne (Economic growth, restructurization and labor market in Poland)*, Wydawnictwo Uniwersytetu Łódzkiego, Łódź.
Vasudeva N., Chien I., 1997, The empirics of economic growth for OECD countries: some new findings, "Economics Letters", vol. 55, p. 425-29.
Welfe, W. (red.), 2001, *Ekonometryczny model wzrostu gospodarczego (Econometric model of economic growth)*, Wydawnictwo Uniwersytetu Łódzkiego, Łódź.
Welfe W., 2005, *A Stylised Simulation Minimodel of Economic Growth*, „Materiały IEiS", no. 1, mimeo.
Welfe W., Florczak W., Sabanty L., 2002, *Kapitał ludzki i jego endogenizacja (Human capital and its endogenization)*, „Przegląd Statystyczny", vol. 49, no. 2.
Woessmann L., 2003, *Specifying Human Capital*, "Journal of Economic Surveys", vol. 17/3, p. 239-270.

Chapter 6

Modelling integrated panel data: an overview[1]

6.1 Introduction

The methods for modelling integrated panel data began with proving that the Dickey-Fuller type statistic for panel data converges to a normal distribution, as a number of cross-sections (I, countries, industries, etc.) and the length of time-series (T) tend to infinity, (see Quah (1990), and Breitung and Meyer (1994)). It should be stressed, however, that their assumptions are exceedingly restrictive. Another major stepping-stone in this area has been limit theory for integrated panel data that Phillips and Moon (1999) developed for large I and T, where the rate condition $I/T \to 0$ is imposed. Philips and Moon considered three variants of deriving asymptotics for such double-indexed processes. Firstly, a sequential limit case, where $T \to \infty$ is followed by $I \to \infty$. Secondly, a diagonal path limit theory, where both indices go to infinity along a specified diagonal path, thus T is determined by a function of I, as $I \to \infty$. Thirdly, a joint limit, where both indices simultaneously go to infinity. It should be stressed that the three cases can have different asymptotics, thus even different normalizations of statistics can be needed to obtain non-degenerate asymptotic distributions. However, under certain conditions formulated by Phillips and Moon (1999), the cases lead to the same asymptotics and the joint limits can be derived by means of straightforward sequential limits. Unlike asymptotics for large I and fixed T, limit distributions derived as described are appropriate for panels with a large number of observations and a moderate number of cross-sections, which are usually investigated at the macro level.

The models of integrated panel data are related to the models of stationary panel data. However, the two types should not be considered as substitutes, because of different assumptions underpinning the stochastic structure of a data-generating process. Other reviews of methods for analysing the integrated panel data can be found, for example, in the papers by Phillips and Moon (2000), Banerjee (1999), Baltagi and Kao (2000).

In the remainder of this chapter, the following notations are used. The operators $tr(\cdot)$ and $rk(\cdot)$ denote the trace and the rank of a matrix. The symbol Ţ stands for convergence in distribution, and the symbol $\xrightarrow{p}$ indicates convergence in probability. The integral $\int_0^1 (d\mathbf{W}(u))^T \mathbf{W}(u)(du)$ is short-handed

[1] The author wishes to thank Professor Władysław Welfe for his suggestion to write this chapter.

as $\int_0^1 (d\mathbf{W})^T \mathbf{W}$, where $\mathbf{W}(u)$ is the multi-dimensional Brownian motion with an identity variance-covariance matrix and $u \in [0;1]$.

6.2 Tests of integration order for panel data

Inference on the integration order of panel data can take advantage of tests making different assumptions about the data-generating process, as well as tests with a hypothesis of a unit root in the null or in the alternative. It should be stressed that the tests assuming a unit root in the null and the tests with a unit root in the alternative complement each other. In turn, assumptions about the stochastic structure of a panel data-generating process should primarily build on statistical reasons supplemented by the knowledge of an economic phenomenon. Moreover, the choice of the appropriate test of integration order is constrained by the rate condition on I and T. Assuming that $\sqrt{I}/T$ goes to zero or constant, the number of cross-sections can even exceed the length of time-series, which is typical of panel data at the micro level. If I/T is assumed to go to zero, the applicability of this test is usually limited to panel data at the macro level. Complying with these assumptions guarantees that the limit distributions will be valid.

Whether a panel data-generating process has a unit root can be verified using the test proposed by Levin, Lin and Chu (2002), henceforth the LLC test. The test assumes that a series is generated by the following process:

$$\Delta y_{it} = \rho_i y_{i,t-1} + \sum_{k=1}^{K_i} \gamma_{ik} \Delta y_{i,t-k} + \boldsymbol{\alpha}_i \mathbf{d}_t + \varepsilon_{it} \tag{6.1}$$

where $i = 1, 2, \ldots, I$, $t = 1, 2, \ldots, T$, $k = 1, 2, \ldots, K_i$, ε_{it} is a stationary variable independent in both time-series and the cross-sections dimension, and $\mathbf{d}_t$ denotes a vector of dummies, therefore $\mathbf{d}_t = 0$, $\mathbf{d}_t = 1$ and $\mathbf{d}_t = [1 \; t]$ signify the model without a constant term, the model with an intercept and the model with an intercept and a linear trend, respectively. The null and alternative hypotheses are as follows:

$$\mathcal{H}_0 : \rho_1 = \rho_2 = \ldots = \rho_I = 0 \tag{6.2a}$$

vs.

$$\mathcal{H}_1 : \rho_1 = \rho_2 = \ldots = \rho_I = \rho < 0 \tag{6.2b}$$

The LLC test statistic is calculated in three steps. First the lag order (K_i) for each cross-section is estimated and model (6.1) is simplified in line with the Frisch-Waugh theorem. Therefore, the maximum lag order is selected for each cross-section depending on the sample size, and then the lag order is determined using a sequential verification of t-type statistics. However, the limit distribution of the Dickey-Fuller type statistic depends on the assumption that the growth rate of K_i is $T^{1/3}$ at the most (see Said and Dickey (1984)). Next, for each cross-section, Δy_{it} and $y_{i,t-1}$ are regressed against $\Delta y_{i,t-k}$ $(k=1,2,\ldots,K_i)$ and deterministic term $\mathbf{d}_t$:

$$\hat{e}_{1it} = \Delta y_{it} - \sum_{k=1}^{K_i} \hat{\gamma}_k \Delta y_{i,t-k} - \hat{\boldsymbol{\alpha}}_i \mathbf{d}_t , \tag{6.3}$$

$$\hat{e}_{2i,t-1} = y_{i,t-1} - \sum_{k=1}^{K_i} \hat{\hat{\gamma}}_k \Delta y_{i,t-k} - \hat{\hat{\boldsymbol{\alpha}}}_i \mathbf{d}_t . \tag{6.4}$$

Because model (6.1) allows the heterogeneity of panel data, the residuals $\hat{e}_{1it}$, $\hat{e}_{2i,t-1}$ are then normalized by the standard deviation of the error term ε_{it}:

$\tilde{e}_{1it} = \hat{e}_{1it} / \hat{\sigma}_{\varepsilon i}$,

$\tilde{e}_{2i,t-1} = \hat{e}_{2i,t-1} / \hat{\sigma}_{\varepsilon i}$,

where the standard deviation of the error term ε_{it} can be calculated as the standard error of the error term in regression of $\hat{e}_{1it}$ against $\hat{e}_{2i,t-1}$:

$$\hat{\sigma}_{\varepsilon_i} = \left((T-K_i-1)^{-1} \sum_{t=K_i+2}^{T} \left(\hat{e}_{1it} - \hat{\rho}_i \hat{e}_{2i,t-1} \right)^2 \right)^{1/2} . \tag{6.5}$$

In the second step, the average ratio of the long-run standard deviation of the error term $\widehat{\varepsilon}_{it}$ in the AR(1) model of y_{it} ($\widehat{\varepsilon}_{it}$ follows a stationary and invertible ARMA process) to the short-run standard deviation of ε_{it} is estimated. The long-run variance $\sigma^2_{0\varepsilon_i} = \lim_{T\to\infty} T^{-1} E\left(\left(\sum_{t=1}^{T} \widehat{\varepsilon}_{it} \right)^2 \right)$ can be derived from a spectral density function at frequency zero. Due to the small sample properties of the test, the following estimator of the long-run variance is used:

$$\hat{\sigma}^2_{0\hat{\varepsilon}_i} = (T-1)^{-1} \sum_{t=2}^{T} \Delta y_{it}^2 + 2(T-1)^{-1} \sum_{j=1}^{J} w(j,J) \sum_{t=j+2}^{T} \Delta y_{it} \Delta y_{i,t-j} , \tag{6.6}$$

where the weights $w(j,J)$ can be calculated by means of the Bartlett kernel, for which $w(j,J) = 1 - j/(J+1)$. It should be stressed that convergence of the long-run variance estimator depends on the growth rate of J, which should grow at rate $T^{1/2}$ at the most. If process (1) generating series Δy_{it} contains deterministic terms (an intercept, a linear trend), then the original series have to be replaced by demeaned series $\Delta y_{it} - \Delta y_{it}$ or detrended series in (6.6). After calculating the estimator of the long-run variance, the ratios of the long-run standard deviation of the error term $\widehat{\varepsilon}_{it}$ to the short-run standard deviation of ε_{it} are estimated for each cross-section:

$$\hat{s}_i = \hat{\sigma}_{0\hat{\varepsilon}_i} / \hat{\sigma}_{\varepsilon_i} , \tag{6.7}$$

and the average ratio is $\hat{S}_I = I^{-1} \sum_{i=1}^{I} \hat{s}_i$.

In the last step, a pooled regression of $\hat{e}_{1it}$ against $\hat{e}_{2i,t-1}$ needs to be run, for observations from all cross-sections:

$$\tilde{e}_{1it} = \hat{\rho}\tilde{e}_{2i,t-1} + \tilde{\varepsilon}_{it}\,. \tag{6.8}$$

The *t*-type statistic of the LLC test for the null hypothesis (6.2a) is given by:

$$t_{\rho} = \frac{\hat{\rho}}{\hat{\sigma}_{\hat{\rho}}}\,. \tag{6.9}$$

The limit distribution of statistic (9) for $\mathbf{d}_t = 0$ is t_{ρ} Ţ $N(0;1)$, as $T \to \infty$, $I \to \infty$ and $\sqrt{I}/T \to 0$.

If the data-generating process contains deterministic terms, then the statistic (6.9) does not converge to a standard normal distribution. Therefore, Levin, Lin and Chu proposed an adjusted *t*-type statistic:

$$t_{\rho}^{*} = \frac{t_{\rho} - I\tilde{T}\hat{S}_I\hat{\sigma}_{\tilde{\varepsilon}}^{-2}\hat{\sigma}_{\hat{\rho}}\mu_{\tilde{T}}}{\sigma_{\tilde{T}}}, \tag{6.10}$$

where $\tilde{T} = T - \overline{K}_i - 1$ is the average number of observations in a cross-section, and $\hat{\sigma}_{\tilde{\varepsilon}}^2$ is the estimator of the error term variance in regression (6.8):

$$\hat{\sigma}_{\tilde{\varepsilon}}^2 = \left(I\tilde{T}\right)^{-1}\sum_{i=1}^{I}\sum_{t=K_i+2}^{T}\left(\tilde{e}_{1it} - \hat{\rho}\tilde{e}_{2i,t-1}\right)^2,$$

$\mu_{\tilde{T}}$ and $\sigma_{\tilde{T}}$ are adjustments of the mean and the standard deviation, respectively, which were tabulated by Levin, Lin and Chu (2002). The limit distribution of statistic (10) is t_{ρ}^{*} Ţ $N(0;1)$, as $T \to \infty$, $I \to \infty$ and $I/T \to 0$.

It should be noted that the t_{ρ} statistic requires a slower growth rate of T than the t_{ρ}^{*} statistic, because their asymptotic distributions are valid for $\sqrt{I}/T \to 0$ and $I/T \to 0$, respectively. However, the t_{ρ} statistic shall not be applied when the data-generating process has deterministic terms, because the limit distribution is not well defined. On the other hand, allowing for the deterministic terms, which are in fact not present, decreases the power of the test.

The results of the Monte Carlo experiments for the t_{ρ}^{*} statistic conducted by Levin, Lin and Chu (2002) show that the small sample properties of the LLC test are satisfactory. The empirical size slightly overestimates the nominal size for $I \leq 10$, but even then, size distortion is usually less than 30 per cent of the nominal size for the model without a constant and several percent for the model with a constant or a linear trend. Moreover, the performance of the LLC test and the Dickey-Fuller test show that the pooling of cross-sections substantially increases the power of a unit root test. It should be stressed, however, that the convergence rate of the Dickey-Fuller type statistic is higher with respect to the time-series dimension than with respect to the number of cross-sections.

The LLC test assumes that series-generating processes in the cross-sections are independent, thus non-existence of the contemporaneous correlation is imposed. Nevertheless, if correlation does exist due to common time effects:

$$\Delta y_{it} = \rho_i y_{i,t-1} + \sum_{k=1}^{K_i} \gamma_k \Delta y_{i,t-k} + \boldsymbol{\alpha}_i \mathbf{d}_t + v_t + \varepsilon_{it}, \tag{6.11}$$

the LLC test can be applied after subtracting the cross-sectional average $\bar{y}_t = I^{-1}\sum_{i=1}^{I} y_{it}$ from observations in each cross-section. In the case of multiple common time effects, the method proposed by Quah and Sargent (1992) can be applied.

The LLC test shall not be applied when ρ_i coefficients are different in the alternative. In such cases, the test proposed by Im, Pesaran and Shin (2003), henceforth the IPS test, is appropriate for inferring on the integration order in panel data. The IPS test assumes that the data are generated by process (6.1), with $K_i = K$ for each cross-section; the null hypothesis is similar to that in the LLC test:

$$\mathcal{H}_0 : \rho_1 = \rho_2 = \ldots = \rho_I = 0, \tag{6.12a}$$

and the alternative hypothesis is as follows:

$$\begin{aligned} &\mathcal{H}_1 : \rho_i < 0 \;\; for \;\; i = 1,2,\ldots,I_1, \quad \rho_i = 0 \;\; for \;\; i = I_1+1, I_1+2,\ldots,I, \\ &\quad \lim_{I\to\infty}\left(I_1/I\right) = \delta, \quad 0 < \delta \le 1. \end{aligned}, \tag{6.12b}$$

Therefore, the IPS test allows different values of ρ_i coefficients in the alternative, as opposed to the LLC test, where the alternative is more restrictive. Moreover, some series may possess a unit root in the alternative.

The IPS test statistic is based on averaging the results for each cross-section. In the first step, the t-type statistics are calculated for the null hypothesis $\rho_i = 0$ for each cross-section, say t_{ρ_i}. Then, the average $\bar{t} = I^{-1}\sum_{i=1}^{I} t_{\rho_i}$ allows writing the statistic of the IPS test:

$$\Upsilon_{\bar{t}} = \frac{\sqrt{I}\left(\bar{t} - E(Z)\right)}{\sqrt{D^2(Z)}}, \tag{6.13}$$

where $E(Z)$ and $D^2(Z)$ denote the mean and the variance of the limit distribution of the Dickey-Fuller test. The moments of the limit depending on the deterministic term were estimated by Nabeyama (1999). The asymptotic distribution of the $\Upsilon_{\bar{t}}$ statistic is $\Upsilon_{\bar{t}} \sim N(0;1)$, as $T\to\infty$, $I\to\infty$ and $I/T \to const$. Therefore, because the growth rates of I and T are expected to be equal, the application of the IPS test is usually limited to the data at the macro level.

The assumption about equal rates of growth of I and T cannot be maintained for the micro-level panel data. However, Im, Pesaran and Shin (2003) proved that for panel data generated by the AR(1) process:

$$\Delta y_{it} = \rho_i y_{i,t-1} + \boldsymbol{\alpha}_i \mathbf{d}_t + \varepsilon_{it}, \tag{6.14}$$

the limit distribution of the $\Upsilon_{\bar{T}}$ statistic for model (6.14) without a linear trend is $\Upsilon_{\bar{T}} \sim N(0;1)$, as $I \to \infty$ and $T > 5$, likewise for the model with a linear trend, because $I \to \infty$ and $T > 6$.

It is possible to infer on the unit root in the unbalanced panel data with different lengths of the time-series in the cross-sections, or with different lag orders K_i, taking advantage of adjusted statistics:

$$\Upsilon_{\bar{T}}^{*} = \frac{\sqrt{I}\left(\bar{t} - I^{-1}\sum_{i=1}^{I} E\left(t_{\rho_i} \middle| \rho_i = 0\right)\right)}{\sqrt{D^2\left(t_{\rho_i} \middle| \rho_i = 0\right)}}. \tag{6.15}$$

The moments of the t_{ρ_i} statistics depend on the form of the deterministic terms in the data-generating process, the sample length T_i, and the lag order K_i for each cross-section. They were calculated using Monte Carlo experiments and tabulated by Im, Pesaran and Shin (2003). The asymptotic distribution of the $\Upsilon_{\bar{T}}^{*}$ statistics is the same as the limit of the $\Upsilon_{\bar{T}}$ statistic. Although they asymptotically provide the same conclusion, their properties in the small samples are different, because the t_{ρ_i} statistic is not pivotal and its expected value depends, for example, on the lag order K_i. Thus, because of the test's small sample properties, the $\Upsilon_{\bar{T}}^{*}$ statistics is recommended for the balanced panel data with small T instead of the $\Upsilon_{\bar{T}}$ statistic.

The results of the Monte Carlo experiments performed by Im, Pesaran and Shin (2003) show that the small sample properties of the IPS test are satisfactory. However, this finding is usually rendered invalid, if the conditions of the IPS test are not fulfilled. For example, an underestimated lag order implies that both the size and power of the test are biased towards zero or unity, depending on the data-generating process (this applies to all unit root tests, including the LLC test). On the other hand, an overestimated lag order causes relatively limited size distortion. Therefore, it is comparatively better to overestimate the lag order than to underestimate it.

In the case of contemporaneous correlation caused by the common time effects the IPS test can be applied, after subtracting the cross-sectional average in the same manner as in the LLC test. However, if the contemporaneous correlation results from more complicated processes, for example:

$$\Delta y_{it} = \rho_i y_{i,t-1} + \sum_{k=1}^{K_i} \gamma_k \Delta y_{i,t-k} + \boldsymbol{\alpha}_i \mathbf{d}_t + \beta_i v_t + \varepsilon_{it}, \tag{6.16}$$

then the size and power of both the LLC test and the IPS test are seriously distorted (see Maddala and Wu (1999)), and these tests shall not be applied to inference on a unit root in the panel data.

Hypotheses (6.12a-b) can be verified using the Fisher test as proposed by Maddala and Wu (1999) and Choi (2001). Then the statistics can be written as:

$$P = -2\sum_{i=1}^{I} \ln p_i, \tag{6.17}$$

where p_i is the marginal significance level (the p-value) of the unit root test for a given cross-section. The probability values p_i are calculated based on any unit root test. Moreover, the data-generating processes and the numbers of observations can differ within the cross-sections. The marginal significance levels can be computed using either asymptotic distributions or exact distributions simulated by means of Monte Carlo experiments. In the first case, the test distribution is given by $P \sim \chi^2_{2I}$, as $T \to \infty$ for any finite I, and in the other case, we have an exact test, with a left-sided acceptance region of the null hypothesis. The Fisher test does not possess the non-degenerate joint limit distribution, because for the true null hypothesis $P \xrightarrow{p} \infty$, as $T \to \infty$ and $I \to \infty$. The marginal significance levels are assumed to be mutually independent and therefore no contemporaneous correlation exists between the stochastic terms in the cross-sections. Nevertheless, this assumption can be easily cancelled, if the small sample distributions of individual statistics are simulated by means of the bootstrap method (see Maddala and Wu (1999)).

On one hand, the results of the Monte Carlo experiments show that size distortion for the Fisher test is smaller than in the LLC test or the IPS test, if the panel data exhibit cross-correlation. On the other hand, when the contemporaneous correlation is not present, the IPS test has somewhat higher power than the Fisher test, and the Fisher test is more powerful than the LLC test. It should be stressed, however, that the LLC test performance is not directly comparable with the small sample properties of the IPS test, or the Fisher test, because the alternatives are different.

It is easy to see that the Fisher test is a more flexible instrument than the other tests. Firstly, in the case of cross-correlation of the stochastic terms plausible inference can be conducted. Secondly, heterogeneity of the data-generating processes in the cross-sections is allowed with respect to both the stochastic terms and deterministic terms. Thirdly, the marginal significance levels can be related to different unit roots tests (or stationarity tests).

For the lack of non-degenerated asymptotic distribution of the Fisher test, as $I \to \infty$, a modification of the Fisher test as proposed by Choi should be considered. Choi (2001) put forward the following statistic:

$$P^* = \frac{I^{-1}\sum_{i=1}^{I}\left(-2\ln p_i - E\left(-2\ln p_i\right)\right)}{\sqrt{D^2\left(-2\ln p_i\right)}} = -I^{-1}\sum_{i=1}^{I}\left(\ln p_i + 1\right), \tag{6.18}$$

where the limit distribution is $P^* \sim N(0;1)$, as $T_i \to \infty$ and $I \to \infty$, if the marginal significance levels are mutually independent.

Choi (2001) proposed two other tests based on the marginal significance levels to verify hypotheses (6.12a-b). The first test uses the following statistic:

$$Z = I^{-1/2}\sum_{i=1}^{I} F_N^{-1}\left(p_i\right), \tag{6.19}$$

where $F_N\left(p_i\right)$ denotes the cumulative distribution function of a standardized normal distribution, and the inverse function generates a variable from this distribution (the marginal significance levels are drawn from the uniform distribution for the $[0;1]$ interval). The other test can be written as:

$$L^* = \sqrt{\frac{3(5I+4)}{\pi^2 I(5I+2)}} \cdot \sum_{i=1}^{I} \ln\left(\frac{p_i}{1-p_i}\right), \tag{6.20}$$

where $\pi = 3.14\ldots$.

The asymptotic distributions of Z and L^* are $Z \sim N(0;1)$ and $L^* \sim t_{5I+4}$ for a finite number of cross-sections, as $T_i \to \infty$. In turn, for $I \to \infty$, both statistics converge to a standardized normal distribution. The rejection region is left-sided for both tests. The small sample study performed by Choi (2001) reveals that the tests based on the marginal significance levels are more powerful than the IPS test, and that the Z test is the most powerful. Moreover, size distortions for the former are comparatively negligible, even under a small number of cross-sections.

A common feature of the unit root tests is their relatively low power in the case of time-series or panel data with a small number of cross-sections, say $I < 10$. This implies that the integration order indicated by the tests exceeds the actual one. Therefore, the stationarity tests represent a valuable supplement to the the unit roots tests. The null hypothesis and the alternative are reversed in these tests, following the stationarity test for time-series proposed by Kwiatkowski et. al (1992). Hadri (2000) proposed to augment the test constructed by Kwiatkowski et al. (1992) for the case of panel data.

Let a series be generated by the following process:

$$y_{it} = r_{it} + \varepsilon_{it} \tag{6.21a}$$

or

$$y_{it} = r_{it} + \beta_i t + \varepsilon_{it} \tag{6.21b}$$

where r_{it} denotes the realization of a random walk, $r_{it} = r_{i,t-1} + u_{it}$, ε_{it} is a weakly stationary variable, $u_{it} \sim N(0;\sigma_{u_i}^2)$, ε_{it} and u_{it} are mutually independent, and independent in both the time dimension and cross-sections, and r_{i0} is an intercept. A straightforward transformation allows writing model (6.21b) as:

$$y_{it} = r_{i0} + \beta_i t + \sum_{t=1}^{T} u_{it} + \varepsilon_{it} = r_{i0} + \beta_i t + e_{it}, \tag{6.22}$$

where $e_{it} = \sum_{t=1}^{T} u_{it} + \varepsilon_{it}$. e_{it} is a weakly stationary variable, only if $\sigma_{u_i}^2 = 0$, thus the null hypothesis of the stationarity test for y_{it} is given as:

$$\mathcal{H}_0 : \sigma_{u_1}^2 = \sigma_{u_2}^2 = \ldots = \sigma_{u_I}^2 = 0, \tag{6.23a}$$

and the alternative is:

$$\mathcal{H}_1 : \sigma_{u_i}^2 > 0 \;\; for \;\; i = 1, 2, \ldots, I_1, \quad \sigma_{u_i}^2 = 0 \;\; for \;\; i = I_1+1, I_1+2, \ldots, I, \quad \lim_{I\to\infty}(I_1/I) = \delta, \quad 0 < \delta \le 1. \tag{6.23b}$$

The Lagrange multiplier statistic is the following:

$$LM = \frac{I^{-1}T^{-2}\sum_{i=1}^{I}\sum_{t=1}^{T} S_{it}^2}{\hat{\sigma}_\varepsilon^2}, \tag{6.24}$$

where S_{it} denotes the partial sum of residuals from the regression of y_{it} against deterministic terms:

$$S_{it} = \sum_{j=1}^{t} \hat{e}_{ij}, \tag{6.25}$$

and $\hat{\sigma}_\varepsilon^2$ is a consistent estimator of the error term variance in equation (6.22):

$$\hat{\sigma}_\varepsilon^2 = \frac{1}{IT}\sum_{i=1}^{I}\sum_{t=1}^{T} \hat{e}_{it}^2. \tag{6.26}$$

The limit distribution of the LM statistic for the model without a linear trend is given as:

$$LM \xrightarrow{p} E\left(\int V_1^2\right), \quad \text{as } T\to\infty, I\to\infty, I/T\to\infty, \tag{6.27}$$

where V_1 denotes the standard Brownian bridge (see Hadri (2000), Kwiatkowski et al. (1992)). For the model with a linear trend:

$$LM \xrightarrow{p} E\left(\int V_2^2\right), \quad \text{as } T\to\infty, I\to\infty, I/T\to\infty, \tag{6.28}$$

where V_2 denotes a so-called second-level Brownian bridge. The rejection region is right-sided.

The standardized LM statistic proposed by Hadri (2000) is as follows:

$$\Psi_{LM} = \frac{\sqrt{I}\left(LM - E\left(\int V_s^2\right)\right)}{\sqrt{D^2\left(\int V_s^2\right)}}, \tag{6.29}$$

where $s=1$ for the model without a linear trend, or $s=2$ for the model with a trend. The asymptotic distribution of Ψ_{LM} is Ψ_{LM} Ţ $N(0;1)$, as $T \to \infty$, $I \to \infty$ and $I/T \to 0$. The rejection region is right-sided. The mean and variance of the limit distribution of the LM statistic for $s=1$ are $\frac{1}{6}$ and $\frac{1}{45}$, respectively, and for $s=2$ they are equal to $\frac{1}{15}$ and $\frac{11}{6300}$ (see Hadri (2000)).

The Hadri test (2000) can also be applied if the temporal dependence exists in the process for generating ε_{it}. For instance, if ε_{it} satisfies regularity conditions as formulated by Phillips and Perron (1988). In this case, the estimator of ε_{it} variance in the denominator of (6.24) has to be replaced by a consistent estimator of the long-run variance.

Hadri and Larsson (2005) proposed the following modified statistic Ψ_{LM}:

$$\Psi^{*}_{LM} = \frac{\sqrt{I}\left(LM - E(LM)\right)}{\sqrt{D^2(LM)}}, \tag{6.30}$$

where for the model without a linear trend:

$$E(LM) = \frac{T+1}{6T}, \quad D^2(LM) = \frac{T^2+1}{20T^2} - \left(\frac{T+1}{6T}\right)^2,$$

and for the model with a linear trend:

$$E(LM) = \frac{T+2}{15T}, \quad D^2(LM) = \frac{(T+2)(13T^2+23)}{2100T^2} - \left(\frac{T+2}{15T}\right)^2.$$

The asymptotic distribution is Ψ^{*}_{LM} Ţ $N(0;1)$, as $I \to \infty$ for a finite T. It should be noted that the test based on the Ψ^{*}_{LM} statistic can be more useful than that founded on the Ψ_{LM} statistic, because the convergence condition is $I \to \infty$. Therefore, the first test can be applied to panel data at the micro level. The results of the Monte Carlo experiments as conducted by Hadri and Larsson (2005) show that size distortions of this test are negligible and smaller than for the test based on the Ψ_{LM} statistic.

The above tests are widely applied instruments enabling inference on the unit roots for panel data. These tests represent the first step in modelling the integrated series. Moreover, they facilitate empirical validation of some economic theories, such as purchasing power parity (see MacDonald (1996), Frankel and Rose (1996) and Maddala and Wu (1999)).

6.3 Modelling panel data integrated of order one with one explained variable in a cross-section.

The modelling of integrated panel data is associated with the phenomenon of spurious regression relating to variables generated by independent random walks. In the case of time-series, this phenomenon implies inconsistency of the least squares dummy variable estimator and falsity of indications of the significance tests. The panel data are also exposed to it, but then the consequences are different. The noise is attenuated by pooling the data which ensures consistency of the least squares dummy variable estimator, but the *t*-ratios remain asymptotically divergent.

Let us consider a bivariate variable $\mathbf{v}_{it} = [\xi_{it}\ \zeta_{it}]$, $i = 1, 2, \dots, I$, $t = 1, 2, \dots, T$, where $E(\mathbf{v}_{it}) = [0\ 0]$, the variance-covariance matrix is:

$$\boldsymbol{\Sigma}_i = \lim_{T\to\infty} T^{-1} \sum_{t=1}^{T} E(\mathbf{v}_{it}^{\mathrm{T}} \mathbf{v}_{it}) = \begin{bmatrix} \sigma_{\xi_i}^2 & \sigma_{\xi_i\zeta_i} \\ \sigma_{\zeta_i\xi_i} & \sigma_{\zeta_i}^2 \end{bmatrix} = \boldsymbol{\Sigma} \quad \text{for each } i,$$

and the long-run variance-covariance matrix is given by:

$$\boldsymbol{\Omega}_i = \lim_{T\to\infty} T^{-1} E\left(\sum_{t=1}^{T} \mathbf{v}_{it}\right)^{\mathrm{T}} \left(\sum_{t=1}^{T} \mathbf{v}_{it}\right) = \begin{bmatrix} \sigma_{0\xi_i}^2 & \sigma_{0\xi_i\zeta_i} \\ \sigma_{0\zeta_i\xi_i} & \sigma_{0\zeta_i}^2 \end{bmatrix} = \boldsymbol{\Omega} \quad \text{for each } i.$$

Let us assume that $\mathbf{v}_{it}$ is independent between cross-sections and satisfies the regularity conditions as formulated by Phillips and Perron (1988). Let $y_{it} = \sum_{j=1}^{t} \xi_{ij}$ and $x_{it} = \sum_{j=1}^{t} \zeta_{ij}$. Then the least squares dummy variable estimator of the following model:

$$y_{it} = \alpha_i + \beta x_{it} + \varepsilon_{it} \tag{6.31}$$

is consistent and $\hat{\beta}_{\mathrm{LS}} \xrightarrow{p} \sigma_{0\xi\zeta} / \sigma_{0\zeta}^2$ (see Entorf (1997), Kao (1999), or Baltagi and Kao (2000)). Moreover, if ξ_{it} and ζ_{it} are mutually independent and independent in both time-series and cross-sections dimension, $E(\xi_{ij}) = 0$, $E(\zeta_{ij}) = 0$, $D^2(\xi_{ij}) = \sigma_\xi^2$, $D^2(\zeta_{ij}) = \sigma_\zeta^2$, $E(\xi_{ij}^4) < \infty$ and $E(\zeta_{ij}^4) < \infty$, then $\hat{\beta}_{\mathrm{LS}} \xrightarrow{p} 0$.

The lack of a non-degenerated limit distribution of the *t*-ratios implies that inference on the existence of a relationship between y_{it} and x_{it} has to be based on the properties of the error term-generating process in model (6.31). On one hand, the regression is spurious when ε_{it} is generated by a random walk. On the other hand, if ε_{it} is a weakly stationary variable, then y_{it} and x_{it} are cointegrated, and model (6.31) shows a long-run static relationship (see Engle and Granger (1987)). However, in the finite samples the least squares dummy variable estimator is biased (see Kao and Chiang (1997)). Moreover, neither the bias-corrected least squares estimator (see Chen et al. (1999)) nor the fully modified ordinary least squares estimator with corrections for endogeneity and serial correlation (see Pedroni (2000)) is appropriate, because the results of the Monte Carlo experiments show that both the estimators are also biased in the finite samples (see Kao and Chiang (1997) and Chen et al. (1999)). The dynamic ordinary least squares estimator has the best small sample properties and its bias is negligible and smaller than the standard deviation of the bias calculated by means of simulation.

The dynamic ordinary least squares estimator (DOLS) of coefficient β in model (6.31) is calculated as a least squares estimator in the following regression:

$$y_{it} = \alpha_i + \beta x_{it} + \sum_{j=-q_1}^{q_2} c_{ij} \Delta x_{i,t+j} + \tilde{\varepsilon}_{it}, \tag{6.32}$$

where q_1 and q_2 are assumed to tend to infinity in such a manner that $q_1^3/I \to 0$, $q_2^3/I \to 0$ and $\sqrt{I} \sum_{|j|>\min(q_1,q_2)} \|c_{ij}\| \to 0$. The asymptotic distribution of the DOLS estimator is a normal one, $E\left(\hat{\beta}_{\mathrm{DOLS}}\right) = \beta$ and $q_1\sqrt{I}T\left(\hat{\beta}_{DOLS} - \beta\right) \xrightarrow{p} N\left(0; 6\sigma_{0\varepsilon.\varsigma} / \sigma_{0\varsigma}^2\right)$, where $\sigma_{0\varepsilon.\varsigma} = \sigma_{0\varepsilon}^2 - \sigma_{0\varepsilon\varsigma}\left(\sigma_{0\varsigma}^2\right)^{-1}\sigma_{0\varsigma\varepsilon}$, (see Kao and Chiang (1997)).

Models (6.31) and (6.32) can be easily augmented to allow for a number of explanatory variables in the cross-sections:

$$y_{it} = \alpha_i + \mathbf{x}_{it}\boldsymbol{\beta} + \varepsilon_{it}, \tag{6.33}$$

where $\mathbf{x}_{it} = \left[x_{1it}\ x_{2it}\ \ldots\ x_{Pit}\right]$ denotes a P-element vector of variables integrated of order one, $\boldsymbol{\beta} = \left[\beta_1\ \beta_2\ \ldots\ \beta_P\right]^{\mathrm{T}}$, the variance-covariance matrices are appropriately extended, and the limit distribution of estimator $\hat{\boldsymbol{\beta}}_{\mathrm{DOLS}}$ is analogous to that in model (6.31). It is also possible to allow for the heterogeneous variance-covariance matrices $\boldsymbol{\Sigma}_i$ and $\boldsymbol{\Omega}_i$ in the cross-sections (see Kao and Chiang (1997)).

Following the application of an unbiased estimator of the long-run relationship we can infer whether the variables are cointegrated, using either the unit roots tests for ε_{it} where the alternative hypothesis assumes the existence of cointegration, or the stationarity test for ε_{it} with the same assumption in the null. The tests based on the Dickey-Fuller type regression for panel data (or the augmented Dickey-Fuller type regression) that Kao (1999) proposed fall into the first category. As for the Dickey-Fuller type tests, ε_{it} is assumed to be generated by a random walk process, so the regression can be written as:

$$\Delta\hat{\varepsilon}_{it} = \rho\hat{\varepsilon}_{i,t-1} + \varsigma_t, \tag{6.34}$$

where $\hat{\varepsilon}_{it}$ denotes residuals from regression (6.31) and ς_t is a weakly stationary variable. The null hypothesis is:

$$\mathcal{H}_0 : \rho_1 = \rho_2 = \ldots = \rho_I = \rho = 0, \tag{6.35a}$$

and the alternative is:

$$\mathcal{H}_1 : \rho_1 = \rho_2 = \ldots = \rho_I = \rho < 0. \tag{6.35b}$$

Kao (1999) proposed the following statistics for hypotheses (6.35a-b):

$$DF_\rho = \frac{\sqrt{I}T\hat{\rho} + 3\sqrt{I}}{\sqrt{51/5}}, \tag{6.36}$$

$$DF_t = \sqrt{5/4}t_\rho + \sqrt{15I/8}, \tag{6.37}$$

$$DF_\rho^* = \frac{\sqrt{I}T\hat{\rho} + \frac{3\sqrt{I}\hat{\sigma}_\varsigma^2}{\hat{\sigma}_{0\varsigma}^2}}{\sqrt{3 + \frac{36\hat{\sigma}_\varsigma^4}{5\hat{\sigma}_{0\varsigma}^4}}}, \tag{6.38}$$

and

$$DF_t^* = \frac{t_\rho + \frac{\sqrt{6I}\hat{\sigma}_\varsigma}{2\hat{\sigma}_{0\varsigma}}}{\sqrt{\frac{\hat{\sigma}_{0\varsigma}^2}{2\hat{\sigma}_\varsigma^2} + \frac{3\hat{\sigma}_\varsigma^2}{10\hat{\sigma}_{0\varsigma}^2}}}, \tag{6.39}$$

where $\hat{\rho}$ is the least squares estimator of coefficient ρ:

$$\hat{\rho} = \frac{\sum_{i=1}^{I}\sum_{t=2}^{T} \Delta\hat{\varepsilon}_{it}\hat{\varepsilon}_{i,t-1}}{\sum_{i=1}^{I}\sum_{t=2}^{T} \hat{\varepsilon}_{i,t-1}^2},$$

t_ρ denotes the t-ratio for the hypothesis (6.35a):

$$t_\rho = \frac{\hat{\rho}}{\hat{\sigma}_\varsigma \Big/ \sqrt{\sum_{i=1}^{I}\sum_{t=2}^{T} \hat{\varepsilon}_{i,t-1}}},$$

and the estimates of variance of ς_{it} and the long-run variance are calculated as follows:

$$\hat{\sigma}_\varsigma^2 = \hat{\sigma}_{\xi_i}^2 - \hat{\sigma}_{\varsigma_i\zeta_i}^2 \hat{\sigma}_{\zeta_i}^2$$

and

$$\hat{\sigma}_{0\varsigma}^2 = \hat{\sigma}_{0\xi_i}^2 - \hat{\sigma}_{0\xi_i\zeta_i}^2 \hat{\sigma}_{0\zeta_i}^2 .$$

The statistics (6.36)-(6.39) converge in the limit to the standardized normal distribution, as $T \to \infty$ and $I \to \infty$. The rejection region is left-sided. It should be noted that if $\hat{\varepsilon}_{i,t-1}$ and ς_{it} are strictly exogenous (which implies the non-existence of serial correlation for ς_{it}), then $\sigma_{\varsigma}^2 = \sigma_{0\varsigma}^2$, and statistics DF_{ρ}^* and DF_t^* simplify to DF_{ρ} and DF_t, respectively.

Apart from the tests based on statistics DF_{ρ}^* and DF_t^*, the augmented Dickey-Fuller type tests can be applied, if ς_{it} is serially correlated. The regression can be written as:

$$\Delta\hat{\varepsilon}_{it} = \tilde{\rho}\hat{\varepsilon}_{i,t-1} + \sum_{j=1}^{K}\gamma_j \Delta\hat{\varepsilon}_{i,t-j} + \tilde{\varsigma}_{it}, \tag{6.40}$$

where $\tilde{\varsigma}_{it}$ is a weakly stationary variable and the lags prevent serial correlation in residuals. The statistic for hypothesis (6.35a) is defined as follows:

$$ADF = \frac{t_{\tilde{\rho}} + \dfrac{\sqrt{6I}\hat{\sigma}_{\varsigma}}{2\hat{\sigma}_{0\varsigma}}}{\sqrt{\dfrac{\hat{\sigma}_{0\varsigma}^2}{2\hat{\sigma}_{\varsigma}^2} + \dfrac{3\hat{\sigma}_{\varsigma}^2}{10\hat{\sigma}_{0\varsigma}^2}}}, \tag{6.41}$$

where $t_{\tilde{\rho}}$ denotes the t-ratio for hypothesis (6.35a) and model (6.40). The ADF statistic converges in the limit to the standardized normal distribution, as $T \to \infty$ and $I \to \infty$. The rejection region is left-sided.

The results of the Monte Carlo experiments performed by Kao (1999) show that size distortions of the tests based on DF_{ρ}^* and DF_t^* statistics are smaller than for DF_{ρ}, DF_t, and ADF. However, the tests based on DF_{ρ} and DF_t statistics are fairly robust against model misspecification. It is worth noting that the asymptotic distributions of the unit roots tests for ε_{it} were derived for a sequential limit case. Therefore, the number of the cross-sections cannot increase as fast as the number of observations in the time-series dimension.

The tests proposed by Kao (1999) assume that the long-run relationship is homogeneous, apart from the intercepts. However, if this assumption is not fulfilled, the results can often lead to a false null hypothesis. Then the model should allow all parameters to vary in the cross-sections:

$$y_{it} = \alpha_i + \mathbf{x}_{it}\boldsymbol{\beta}_i + \varepsilon_{it}, \tag{6.42}$$

where $\boldsymbol{\beta}_i = [\beta_{1i}\ \beta_{2i} \ldots \beta_{Pi}]^T$; it is additionally assumed that the variables in the distinct cross-sections are not contemporaneously correlated, and that the variance-covariance matrices $\boldsymbol{\Sigma}_i$ and $\boldsymbol{\Omega}_i$ are allowed to vary with i. Therefore, model (6.42) allows for the heterogeneity of panel data to a large extent.

Pedroni (2000) introduced seven unit roots tests for ε_{it} that can be divided into two sets. First, he proposed a method for deriving the statistics of the tests for panel data where the means of nominators and denominators of statistics (of the Dickey-Fuller and Phillips and Perron (1988) type) were calculated separately for each cross-section, and the ratios of these means were standardized by correcting the expected values and standard deviations Thus the statistics converged to the standardized normal distribution. Kao formulated identical hypotheses for his tests.

The second set of tests is based on averaging the statistics in the cross-sections, as in the IPS test, and on the standardization of statistics. The null hypothesis for this set of tests is given as:

$$\mathcal{H}_0 : \rho_1 = \rho_2 = \ldots = \rho_I = 0, \tag{6.43a}$$

and the alternative is:

$$\begin{aligned}&\mathcal{H}_1 : \rho_i < 0 \ \ for \ \ i = 1,2,\ldots,I_1, \quad \rho_i = 0 \ \ for \ \ i = I_1+1, I_1+2,\ldots,I,\\ &\quad \lim_{I\to\infty}(I_1/I) = \delta, \quad 0 < \delta \le 1,\end{aligned} \tag{6.43b}$$

where, depending on the test, ρ_i is a parameter in model (6.34) or (6.40) (see Pedroni (2000)).

The long-run relationship can also be validated using a stationarity test for ε_{it}, where the null hypothesis assumes that cointegration occurs and the alternative is a spurious regression, as in the test proposed by McCoskey and Kao (1998). Let us consider model (6.42) and a process for generating ε_{it}:

$$\begin{aligned}&\varepsilon_{it} = r_{it} + v_{it},\\ &r_{it} = r_{i,t-1} + u_{it},\end{aligned} \tag{6.44}$$

where v_{it} is a weakly stationary variable, $u_{it} \sim N\left(0;\sigma^2_{u_i}\right)$, v_{it} and u_{it} are mutually independent and independent in the time dimension and cross-sections. The null hypothesis is:

$$\mathcal{H}_0 : \sigma^2_{u_1} = \sigma^2_{u_2} = \ldots = \sigma^2_{u_I} = 0, \tag{6.45a}$$

and the alternative is:

$$\begin{aligned}&\mathcal{H}_1 : \sigma^2_{u_i} > 0 \ \ for \ \ i = 1,2,\ldots,I_1, \quad \sigma^2_{u_i} = 0 \ \ for \ \ i = I_1+1, I_1+2,\ldots,I,\\ &\quad \lim_{I\to\infty}(I_1/I) = \delta, \quad 0 < \delta \le 1.\end{aligned} \tag{6.45b}$$

The statistic of the test proposed by McCoskey and Kao (1998) can be written as:

$$\overline{LM} = \frac{I^{-1}T^{-2}\sum_{i=1}^{I}\sum_{t=1}^{T}S^2_{it}}{\hat{\sigma}_{0v.\zeta}}, \tag{6.46}$$

where S_{it} is the partial sum process of residuals from model (6.42) regressed by means of an unbiased method such as the DOLS or the FMOLS estimator, and $\hat{\sigma}_{0v.\zeta} = \hat{\sigma}^2_{0v} - \hat{\mathbf{\Omega}}_{0v\zeta}\hat{\mathbf{\Omega}}^{-2}_{0\zeta}\hat{\mathbf{\Omega}}_{0\zeta\zeta}$. In turn, the standardized statistic $\overline{LM}$ is of the form:

$$\Psi_{\overline{LM}} = \frac{\sqrt{I}\left(\overline{LM} - E\left(\int \mathrm{V}^2_s\right)\right)}{\sqrt{D^2\left(\int \mathrm{V}^2_s\right)}}. \tag{6.47}$$

The $\Psi_{\overline{LM}}$ statistic converges in distribution to the standardized normal distribution, as $T \to \infty$, $I \to \infty$ and $I/T \to \infty$. The rejection region of the null is right-sided.

The tests for panel data cointegration help to validate relationships such as the influence of foreign R&D investments on country's productivity (see Kao et al. (1999), or Lee (2005)), the contribution of expanding financial markets to economic growth (Pelgrin and Schich (2002)), or the influence of the urbanization rate on GDP (McCoskey and Kao (1999)).

6.4 Modelling panel data integrated of order one with many explained variables in the cross-section

When the non-stationary panel data with one explained variable in the cross-section are being modelled, the contemporaneous correlations between variables in the distinct cross-sections are rejected. However, if the variance-covariance matrix of the multidimensional cross-sectional stochastic term is not block diagonal, then the approach based on a single equation can easily lead to false conclusions. This structure of the variance-covariance matrix can be often deduced from economic theories (e.g., theories explaining exchange rates or import/export prices of tradables), or from basic statistical analysis.

Models of integrated panel data with one explained variable in the cross-section omit decomposition of the cointegration space into structural long-run relationships. The impacts of the stochastic attractors between the cross-sections cannot be identified, either. This situation makes the intensively developing methods within the multivariate cointegration analysis of panel data especially interesting.

Let $\mathbf{y}_{it} = \left[y_{1it} \; y_{2it} \; \dots \; y_{Pit} \right]$ denote a P-element vector of variables integrated of order one ($p = 1, 2, \dots, P$) in the cross-section i ($i = 1, 2, \dots, I$) and in time t ($t = 1, 2, \dots, T$), and $\mathbf{y}_t = \left[\mathbf{y}_{1t} \; \mathbf{y}_{2t} \; \dots \; \mathbf{y}_{It} \right]$ signify an IP-element vector of variables in the cross-sections in time t. The panel vector error correction model can be written as:

$$\Delta \mathbf{y}_t = \mathbf{y}_{t-1} \mathbf{\Pi} + \sum_{k=1}^{K} \Delta \mathbf{y}_{t-k} \mathbf{\Gamma}_k + \boldsymbol{\varepsilon}_t, \qquad (6.48a)$$

where $\mathbf{\Pi}$ and $\mathbf{\Gamma}_k$ ($k = 1, 2, \dots, K$) are $IP \times IP$ matrices of parameters, $\Delta \mathbf{y}_t$, $\Delta \mathbf{y}_{t-1}$, $\Delta \mathbf{y}_{t-k}$ are IP-element vectors of the lagged first differences, $\boldsymbol{\varepsilon}_t = \left[\varepsilon_{11t} \dots \varepsilon_{P1t} \; \varepsilon_{12t} \dots \varepsilon_{P2t} \dots \varepsilon_{1It} \dots \varepsilon_{PIt} \right]$ and $\boldsymbol{\varepsilon}_t \sim N_{IP}\left(\mathbf{0}; \mathbf{\Omega} \right)$. For the deterministic terms in a data-generating process, model (6.48a) should be augmented as follows:

$$\Delta \mathbf{y}_t = \mathbf{y}_{t-1} \mathbf{\Pi} + \sum_{k=1}^{K} \Delta \mathbf{y}_{t-k} \mathbf{\Gamma}_k + \mathbf{d}_t \mathbf{\Phi} + \boldsymbol{\varepsilon}_t, \qquad (6.48b)$$

where $\mathbf{d}_t$ is the vector of dummies and $\mathbf{\Phi}$ is the matrix of parameters.

A standard analysis of non-stationary panel data is based on a block-diagonal version of model (6.48):

$$\Delta\mathbf{y}_t = \begin{bmatrix}\mathbf{y}_{1t} & \mathbf{y}_{2t} & \cdots & \mathbf{y}_{It}\end{bmatrix}\begin{bmatrix}\mathbf{\Pi}_{11} & \mathbf{0} & \cdots & \mathbf{0}\\ \mathbf{0} & \mathbf{\Pi}_{22} & \cdots & \mathbf{0}\\ \vdots & \vdots & \ddots & \vdots\\ \mathbf{0} & \mathbf{0} & \cdots & \mathbf{\Pi}_{II}\end{bmatrix}+$$

$$+\sum_{k=1}^{K}\begin{bmatrix}\Delta\mathbf{y}_{1,t-k} & \Delta\mathbf{y}_{2,t-k} & \cdots & \Delta\mathbf{y}_{I,t-k}\end{bmatrix}\begin{bmatrix}\mathbf{\Gamma}_{11,k} & \mathbf{0} & \cdots & \mathbf{0}\\ \mathbf{0} & \mathbf{\Gamma}_{22,k} & \cdots & \mathbf{0}\\ \vdots & \vdots & \ddots & \vdots\\ \mathbf{0} & \mathbf{0} & \cdots & \mathbf{\Gamma}_{II,k}\end{bmatrix}+ \tag{6.49}$$

$$+\begin{bmatrix}\boldsymbol{\varepsilon}_{1t} & \boldsymbol{\varepsilon}_{2t} & \cdots & \boldsymbol{\varepsilon}_{It}\end{bmatrix} = \mathbf{y}_t\mathbf{\Pi}_A + \sum_{k=1}^{K}\Delta\mathbf{y}_{t-k}\mathbf{\Gamma}_{A,k} + \boldsymbol{\varepsilon}_t\,,$$

where $\mathbf{\Pi}_{ii}$ and $\mathbf{\Gamma}_{ii,k}$ are $P\times P$ matrices of parameters, and the variance-covariance matrix of the error term is:

$$\mathbf{\Omega} = \begin{bmatrix}\mathbf{\Omega}_{11} & \mathbf{0} & \cdots & \mathbf{0}\\ \mathbf{0} & \mathbf{\Omega}_{22} & \cdots & \mathbf{0}\\ \vdots & \vdots & \ddots & \vdots\\ \mathbf{0} & \mathbf{0} & \cdots & \mathbf{\Omega}_{II}\end{bmatrix}. \tag{6.50}$$

It should be noted that each cross-section is modelled independently of the others, thus the number of the long-run relationships is calculated separately and the parameters in the cross-sections are not related. This means that the following models of subsystems are being considered:

$$\Delta\mathbf{y}_{it} = \mathbf{y}_{i,t-1}\mathbf{\Pi}_{ii} + \sum_{k=1}^{K}\Delta\mathbf{y}_{i,t-k}\mathbf{\Gamma}_{ii,k} + \boldsymbol{\varepsilon}_{it}\,. \tag{6.51}$$

If $\mathbf{\Pi}_{ii}$ is a reduced rank matrix, then it can be decomposed as $\mathbf{\Pi}_{ii} = \mathbf{B}_{ii}\mathbf{A}_{ii}^{\mathrm{T}}$, so that $\mathbf{A}_{ii}$ and $\mathbf{B}_{ii}$ are $P\times R_i$ matrices of full rank, $\mathbf{A}_{ii}$ denotes loadings matrices, and $\mathbf{B}_{ii}$ consists of R_i cointegration vectors. Therefore, a reduced rank of $\mathbf{\Pi}_{ii}$ implies that the variables cointegrate. The parameters are usually estimated by means of the maximum likelihood method and the Frisch-Waugh theorem, i.e. by the reduced rank regression (see Johansen (1988)).

The cointegration rank (rank of $\mathbf{\Pi}_{ii}$) of each subsystem is usually estimated using a so-called trace test, for which the null hypothesis is defined as:

$\mathcal{H}_0 : rk(\mathbf{\Pi}_{ii}) \le R_i$,

and the alternative is:

$\mathcal{H}_1 : rk\left(\mathbf{\Pi}_{ii}\right) = P$.

The statistic of the trace test is the likelihood ratio:

$$LR_{iT}\left(\mathcal{H}_0\left(R_i\right) \middle| \mathcal{H}_1\left(P\right)\right) = -T \sum_{j=R_i+1}^{P} \ln\left(1-\hat{\lambda}_{ij}\right). \tag{6.52}$$

where $\hat{\lambda}_{ij}$ denotes j^{th} eigenvalue (in a descending order) derived from the reduced rank regression of model (6.51). The limit distribution of the trace test statistic is the following (see Johansen (1996)):

$$\begin{aligned} & LR_{iT}\left(\mathcal{H}_0\left(R_i\right) \middle| \mathcal{H}_1\left(P\right)\right) \Rightarrow \; Z_{P-R_i} \equiv \\ & \equiv tr\left(\int_0^1 \left(d\mathbf{W}\right)^T \mathbf{F}\left[\int_0^1 \mathbf{F}^T \mathbf{F} du\right]^{-1} \int_0^1 \mathbf{F}^T \left(d\mathbf{W}\right)\right), \quad as\ T \to \infty, \end{aligned} \tag{6.53}$$

where $\mathbf{W}$ is the $P-R_i$ - dimensional Brownian motion with an identity variance-covariance matrix, (see, for instance, Spanos (1999)), $u \in [0;1]$, and $\mathbf{F}$ is the Brownian motion corrected for a constant or a linear trend (if $\mathbf{\Phi} = \mathbf{0}$, then $\mathbf{F} = \mathbf{W}$).

It is also possible to infer on the cointegration rank using the so-called maximum eigenvalue test, for which the null hypothesis is given as:

$\mathcal{H}_0 : rk\left(\mathbf{\Pi}_{ii}\right) \le R_i$

and the alternative is:

$\mathcal{H}_1 : rk\left(\mathbf{\Pi}_{ii}\right) \le R_i + 1$

The statistic of the maximum eigenvalue test is the likelihood ratio:

$$LR_{iT}^{\max}\left(\mathcal{H}_0\left(R_i\right) \middle| \mathcal{H}_1\left(R_i+1\right)\right) = -T \ln\left(1-\hat{\lambda}_{i,r+1}\right). \tag{6.54}$$

The asymptotic distribution of the maximum eigenvalue test statistic is (see Johansen (1996)):

$$\begin{aligned} & LR_{iT}^{\max}\left(\mathcal{H}_0\left(R_i\right) \middle| \mathcal{H}_1\left(R_i+1\right)\right) \Rightarrow \; Z_{P-R_i}^{\max} \equiv \\ & \equiv \lambda_{\max}\left(\int_0^1 \left(d\mathbf{W}\right)^T \mathbf{F}\left[\int_0^1 \mathbf{F}^T \mathbf{F} du\right]^{-1} \int_0^1 \mathbf{F}^T \left(d\mathbf{W}\right)\right), \quad as\ T \to \infty. \end{aligned} \tag{6.55}$$

The above-mentioned model and the methods of inference on the cointegration rank for panel data integrated of order one represent I separate cointegration analyses, and the information from the cross-section dimension does not affect the results. It is therefore sensible to estimate a common cointegration rank for all cross-sections as Larsson, Lyhagen and Löthgren (2001) proposed; then the hypotheses are as follows:

$$\mathcal{H}_0 : rk\left(\mathbf{\Pi}_{ii}\right) = R_i \le R \quad for\ \ i = 1,\ 2,\ \ldots,\ I, \tag{6.56a}$$

versus the alternative:

$$\mathcal{H}_1 : rk\left(\mathbf{\Pi}_{ii}\right) = P \quad for\ \ i = 1,\ 2,\ \ldots,\ I. \tag{6.56b}$$

The statistic is defined as the average of $LR_{iT}\left(\mathcal{H}_0\left(R_i\right)|\mathcal{H}_1\left(P\right)\right)$ statistics for individual cross-sections:

$$\overline{LR}_{IT}\left(\mathcal{H}_0\left(R_i\right)|\mathcal{H}_1\left(P\right)\right)=\frac{1}{I}\sum_{i=1}^{I}LR_{iT}\left(\mathcal{H}_0\left(R_i\right)|\mathcal{H}_1\left(P\right)\right). \tag{6.57}$$

Calculation of the common cointegration rank R enables the following decomposition of matrix $\mathbf{\Pi}$:

$$\mathbf{\Pi}=\mathbf{\Pi}_B=\begin{bmatrix}\mathbf{B}_{11}\mathbf{A}_{11}^T & \mathbf{0} & \cdots & \mathbf{0}\\ \mathbf{0} & \mathbf{B}_{22}\mathbf{A}_{22}^T & \cdots & \mathbf{0}\\ \vdots & \vdots & \ddots & \vdots\\ \mathbf{0} & \mathbf{0} & \cdots & \mathbf{B}_{II}\mathbf{A}_{II}^T\end{bmatrix}. \tag{6.58}$$

where each $\mathbf{B}_{ii}$ is a full rank matrix with R cointegration vectors.

Larsson et al. (2001) proposed to conduct an inference procedure based on the standardized $\overline{LR}_{IT}\left(\mathcal{H}_0\left(R_i\right)|\mathcal{H}_1\left(P\right)\right)$ statistic:

$$\Upsilon_{\overline{LR}_{IT}}\left(\mathcal{H}_0\left(R_i\right)|\mathcal{H}_1\left(P\right)\right)=\frac{\sqrt{I}\left(\overline{LR}_{IT}-E\left(\mathrm{Z}_{P-R_i}\right)\right)}{\sqrt{D^2\left(\mathrm{Z}_{P-R_i}\right)}}, \tag{6.59}$$

where $E\left(\mathrm{Z}_{P-R_i}\right)$ and $D^2\left(\mathrm{Z}_{P-R_i}\right)$ denote a mean and a variance of the limit distribution of the $LR_{iT}\left(\mathcal{H}_0\left(R_i\right)|\mathcal{H}_1\left(P\right)\right)$ statistic given by (6.53).The asymptotic distribution is

$$\Upsilon_{\overline{LR}_{IT}}\left(\mathcal{H}_0\left(R_i\right)|\mathcal{H}_1\left(P\right)\right)\Rightarrow N(0;1) \quad \text{as } T\to\infty,\ I\to\infty \text{ and } \sqrt{I}/T\to 0. \tag{6.60}$$

The rejection region of the null is right-sided.

It should be underlined that the $\Upsilon_{\overline{LR}_{IT}}\left(\mathcal{H}_0\left(R_i\right)|\mathcal{H}_1\left(P\right)\right)$ statistic converges to the standardized normal distribution, if the rate condition $\sqrt{I}/T\to 0$ is preserved. The Monte Carlo experiments performed by Larsson et al. (2001) show that enlargement of I with respect to fixed T increases the test size towards unity. It can be concluded that the size distortions are relatively small, if $\sqrt{I}/T\le 0.05$.

Karaman (2008) proposed to estimate the common cointegration rank for model (6.49) with the variance-covariance matrix of the error term as given by (6.50) using also an average of statistics $LR_{iT}^{\max}\left(\mathcal{H}_0\left(R_i\right)|\mathcal{H}_1\left(R_i+1\right)\right)$:

$$\overline{LR}_{IT}^{\max}\left(\mathcal{H}_0\left(R_i\right)|\mathcal{H}_1\left(R_i+1\right)\right)=\frac{1}{I}\sum_{i=1}^{I}LR_{IT}^{\max}\left(\mathcal{H}_0\left(R_i\right)|\mathcal{H}_1\left(R_i+1\right)\right). \tag{6.61}$$

The hypotheses of the test are given as:

$$\mathcal{H}_0: rk\left(\mathbf{\Pi}_{ii}\right)=R_i\le R \quad \textit{for } i=1,\ 2,\ \ldots,\ I, \tag{62a}$$

versus the alternative:

$$\mathcal{H}_0 : rk(\mathbf{\Pi}_{ii}) = R_i \leq R+1 \quad for\ i=1,\ 2,\ ...,\ I. \tag{62b}$$

The standardized $\overline{LR}_{IT}^{\max}\left(\mathcal{H}_0(R_i)|\mathcal{H}_1(R_i+1)\right)$ statistic is

$$\Upsilon_{\overline{LR}_{IT}^{\max}}\left(\mathcal{H}_0(R_i)|\mathcal{H}_1(R_i+1)\right) = \frac{\sqrt{I}\left(\overline{LR}_{IT}^{\max} - E\left(Z_{P-R_i}^{\max}\right)\right)}{\sqrt{D^2\left(Z_{P-R_i}^{\max}\right)}}, \tag{6.63}$$

where $E\left(Z_{P-R_i}^{\max}\right)$ and $D^2\left(Z_{P-R_i}^{\max}\right)$ are respectively a mean and a variance of the limit distribution of the $LR_{iT}^{\max}\left(\mathcal{H}_0(R_i)|\mathcal{H}_1(R_i+1)\right)$ statistic given in (6.55).The asymptotic distribution is

$$\Upsilon_{\overline{LR}_{IT}^{\max}}\left(\mathcal{H}_0(R_i)|\mathcal{H}_1(R_i+1)\right) \Rightarrow N(0;1) \quad \text{as } T\to\infty,\ I\to\infty \text{ and } \sqrt{I}/T\to 0. \tag{6.64}$$

The rejection region of the null is right-sided.

The results of the Monte Carlo experiments conducted by Karaman (2008) show that size distortions of the tests $\Upsilon_{\overline{LR}_{IT}}\left(\mathcal{H}_0(R_i)|\mathcal{H}_1(P)\right)$ and $\Upsilon_{\overline{LR}_{IT}^{\max}}\left(\mathcal{H}_0(R_i)|\mathcal{H}_1(R_i+1)\right)$ are comparatively small, and that the power of the tests is generally satisfactory, especially when the short-run dynamics is high. However, if many cointegrating vectors exist, the test based on the $\Upsilon_{\overline{LR}_{IT}}\left(\mathcal{H}_0(R_i)|\mathcal{H}_1(P)\right)$ statistic is more powerful than that based on the $\Upsilon_{\overline{LR}_{IT}^{\max}}\left(\mathcal{H}_0(R_i)|\mathcal{H}_1(R_i+1)\right)$ (see similar results for the case of time-series data in Lütkepohl et al. (2001)).

Model (6.49) with a determined common cointegration rank allows inferring on the common overidentified cointegrating vectors for all cross-sections: $\mathcal{H}_0 : \mathbf{B}_{11} = \mathbf{B}_{22} = \ldots = \mathbf{B}_{II}$, so if the null is true, matrix $\mathbf{\Pi}$ can be written as:

$$\mathbf{\Pi} = \mathbf{\Pi}_C = \begin{bmatrix} \mathbf{B}_{II}\mathbf{A}_{11}^T & \mathbf{0} & \cdots & \mathbf{0} \\ \mathbf{0} & \mathbf{B}_{II}\mathbf{A}_{22}^T & \cdots & \mathbf{0} \\ \vdots & \vdots & \ddots & \vdots \\ \mathbf{0} & \mathbf{0} & \cdots & \mathbf{B}_{II}\mathbf{A}_{II}^T \end{bmatrix}. \tag{6.65}$$

Considering that the likelihood ratio test of overidentifying restrictions is χ^2 distributed for each cross-section and that individual statistics are independent, the sum of these statistics is also χ^2 distributed, with the number of degrees of freedom for such a test being equal to the sum of degrees of freedom for each cross-section.

The block-diagonal structure of the variance-covariance matrix $\mathbf{\Omega}$ should be carefully verified, as this assumption of matrix construction implies that the variables in particular cross-sections are mutually independent, and it is not fulfilled when economies at different levels of aggregation (countries, industries, regions, etc.) are being modelled. Then non-stationary panel data should be analysed based on model (6.49) with the following variance-covariance matrix of the error term:

$$\mathbf{\Omega} = \begin{bmatrix} \mathbf{\Omega}_{11} & \mathbf{\Omega}_{12} & \cdots & \mathbf{\Omega}_{1I} \\ \mathbf{\Omega}_{21} & \mathbf{\Omega}_{22} & \cdots & \mathbf{\Omega}_{2I} \\ \vdots & \vdots & \ddots & \vdots \\ \mathbf{\Omega}_{I1} & \mathbf{\Omega}_{I2} & \cdots & \mathbf{\Omega}_{II} \end{bmatrix}, \tag{6.66}$$

as proposed by Groen and Kleibergen (2003).

The asymptotic distribution of the likelihood ratio statistic for hypotheses (56a-b) is as follows:

$$LR_{IT}\left(\mathcal{H}_0\left(\mathbf{\Pi}_B\right)\middle|\mathcal{H}_1\left(\mathbf{\Pi}_A\right)\right) \text{Ţ}$$

$$\text{Ţ} \sum_{i=1}^{I} \operatorname{tr}\left(\int_0^1 \left(d\mathbf{W}_i\right)^T \mathbf{F}_i \left[\int_0^1 \mathbf{F}_i^T \mathbf{F}_i du \right]^{-1} \int_0^1 \mathbf{F}_i^T \left(d\mathbf{W}_i\right) \right) \quad \text{as } T \to \infty, \text{ for fixed } I, \tag{6.67}$$

where $\mathbf{W}_i$ is the $P - R_i$-dimensional Brownian motion with an identity variance-covariance matrix for the cross-section i, and $\mathbf{F}_i$ is the Brownian motion corrected for a constant or a linear trend (if $\mathbf{\Phi} = \mathbf{0}$, then $\forall_i \mathbf{F}_i = \mathbf{W}_i$). It should be noted that the limit distribution of the $LR_{IT}\left(\mathcal{H}_0\left(\mathbf{\Pi}_B\right)\middle|\mathcal{H}_1\left(\mathbf{\Pi}_A\right)\right)$ statistic is valid for the fixed number of cross-sections. Therefore, this test is basically applicable only if the number of observations reasonably exceeds the number of cross-sections.

For the $LR_{IT}\left(\mathcal{H}_0\left(\mathbf{\Pi}_B\right)\middle|\mathcal{H}_1\left(\mathbf{\Pi}_A\right)\right)$ statistic to be calculated, the maximum likelihood estimates of the parameters of model (6.49) with the variance-covariance matrix (6.66) have to be computed. However, the reduced rank regression cannot be applied to model (6.49). Therefore, it is necessary to estimate the parameters using the optimization or switching methods. Groen and Kleibergen (2003) proposed to use an iterative generalized method of moments in order to obtain the maximum likelihood estimates.

The standardized $LR_{IT}\left(\mathcal{H}_0\left(\mathbf{\Pi}_B\right)\middle|\mathcal{H}_1\left(\mathbf{\Pi}_A\right)\right)$ statistic can be written as:

$$\Upsilon_{LR_{IT}}\left(\mathcal{H}_0\left(\mathbf{\Pi}_B\right)\middle|\mathcal{H}_1\left(\mathbf{\Pi}_A\right)\right) = \frac{LR_{IT}\left(\mathcal{H}_0\left(\mathbf{\Pi}_B\right)\middle|\mathcal{H}_1\left(\mathbf{\Pi}_A\right)\right)/I - E\left(\mathrm{Z}_{P-R_i}\right)}{\sqrt{D^2\left(\mathrm{Z}_{P-R_i}\right)/I}}. \tag{6.68}$$

The limit distribution of the $\Upsilon_{LR_{IT}}\left(\mathcal{H}_0\left(\mathbf{\Pi}_B\right)\middle|\mathcal{H}_1\left(\mathbf{\Pi}_A\right)\right)$ statistic is:

$$\Upsilon_{LR_{IT}}\left(\mathcal{H}_0\left(\mathbf{\Pi}_B\right)\middle|\mathcal{H}_1\left(\mathbf{\Pi}_A\right)\right) \text{Ţ } N(0;1) \quad \text{as } T \to \infty,\ I \to \infty \text{ and } I/T \to 0. \tag{6.69}$$

It should be noted that the limit distribution of the $\Upsilon_{LR_{IT}}\left(\mathcal{H}_0\left(\mathbf{\Pi}_B\right)\middle|\mathcal{H}_1\left(\mathbf{\Pi}_A\right)\right)$ statistics depends on the rate condition. Therefore, the test shall be applied to panel data with T greater than I.

Model (6.49) with the variance-covariance matrix (6.66) and the common cointegration rank R allows inferring on the common, exactly identified cointegrating vectors for all cross-sections: $\mathcal{H}_0 : \mathbf{B}_{11} = \mathbf{B}_{22} = \ldots = \mathbf{B}_{II}$. The asymptotic distribution of statistic of the test for common cointegration space is given by:

$$LR_{IT}\left(\mathcal{H}_0\left(\mathbf{\Pi}_C\right)|\mathcal{H}_1\left(\mathbf{\Pi}_B\right)\right) \text{ Ţ } \chi^2_{(I-1)R(P-R)} \quad \text{as } T \to \infty, \text{ for fixed } I. \tag{6.70}$$

Moreover, as it holds that:

$$\begin{aligned} &LR_{IT}\left(\mathcal{H}_0\left(\mathbf{\Pi}_C\right)|\mathcal{H}_1\left(\mathbf{\Pi}_A\right)\right) = \\ &= LR_{IT}\left(\mathcal{H}_0\left(\mathbf{\Pi}_C\right)|\mathcal{H}_1\left(\mathbf{\Pi}_B\right)\right) + LR_{IT}\left(\mathcal{H}_0\left(\mathbf{\Pi}_B\right)|\mathcal{H}_1\left(\mathbf{\Pi}_A\right)\right), \end{aligned} \tag{6.71}$$

the likelihood ratio statistic for the null $\mathcal{H}_0 : \mathbf{B}_{11} = \mathbf{B}_{22} = \ldots = \mathbf{B}_{II}$ against the alternative $\mathcal{H}_1 : \mathbf{\Pi} = \mathbf{\Pi}_A$ converges to the following convolution of the Dickey-Fuller type distribution and the χ^2 distribution:

$$\begin{aligned} &LR_{IT}\left(\mathcal{H}_0\left(\mathbf{\Pi}_C\right)|\mathcal{H}_1\left(\mathbf{\Pi}_A\right)\right) \text{ Ţ } \chi^2_{(I-1)R(P-R)} + \\ &+ \sum_{i=1}^{I} \operatorname{tr}\left(\int_0^1 \left(d\mathbf{W}_i\right)^T \mathbf{F}_i \left[\int_0^1 \mathbf{F}_i^T \mathbf{F}_i du\right]^{-1} \int_0^1 \mathbf{F}_i^T \left(d\mathbf{W}_i\right) \right) \quad \text{as } T \to \infty, \text{ for fixed } I. \end{aligned} \tag{6.72}$$

The standardized $LR_{IT}\left(\mathcal{H}_0\left(\mathbf{\Pi}_C\right)|\mathcal{H}_1\left(\mathbf{\Pi}_A\right)\right)$ statistic can be written as:

$$\begin{aligned} &\Upsilon_{LR_{IT}}\left(\mathcal{H}_0\left(\mathbf{\Pi}_C\right)|\mathcal{H}_1\left(\mathbf{\Pi}_A\right)\right) = \\ &= \frac{LR_{IT}\left(\mathcal{H}_0\left(\mathbf{\Pi}_C\right)|\mathcal{H}_1\left(\mathbf{\Pi}_A\right)\right)/I - R(P-R)(I-1)/I - E\left(\mathrm{Z}_{P-R_i}\right)}{\sqrt{\left(2(I-1)R(P-R) + D^2\left(\mathrm{Z}_{P-R_i}\right)\right)/I}}. \end{aligned} \tag{6.73}$$

The limit distribution of the statistic $LR_{IT}\left(\mathcal{H}_0\left(\mathbf{\Pi}_C\right)|\mathcal{H}_1\left(\mathbf{\Pi}_A\right)\right)$ is as follows:

$$\Upsilon_{LR_{IT}}\left(\mathcal{H}_0\left(\mathbf{\Pi}_C\right)|\mathcal{H}_1\left(\mathbf{\Pi}_A\right)\right) \text{ Ţ } N(0;1) \quad \text{as } T \to \infty,\ I \to \infty \text{ and } I/T \to 0. \tag{6.74}$$

The results of the Monte Carlo experiments performed by Groen and Kleibergen (2003) show that the likelihood ratio test for the common cointegration rank is much more powerful in the case of the common cointegration space than the likelihood ratio test for an individual cross-section. Moreover, if the variance-covariance matrix $\mathbf{\Omega}$ is not block-diagonal, then the empirical size of the likelihood ratio statistic (6.57) is distinct from the nominal size even in the limit.

The assumption about the block-diagonal structure of model (6.48) (see model (6.49)) needs to be verified. A number of economic theories suggest that some variables in one cross-section depend on the long-run relationships in other cross-sections. If so, the matrix of the long-run multipliers $\mathbf{\Pi}$ cannot have the block-diagonal structure as in $\mathbf{\Pi}_A$. Moreover, if the higher-order cross-correlations of variables in distinct cross-sections are non-zero, then the assumption of the block-diagonal structure of the $\mathbf{\Gamma}_k$ matrices may not be fulfilled, either.

Let us consider the following augmentation of model (6.49):

$$\Delta \mathbf{y}_t = \begin{bmatrix} \mathbf{y}_{1t} & \mathbf{y}_{2t} & \cdots & \mathbf{y}_{It} \end{bmatrix} \begin{bmatrix} \mathbf{\Pi}_{11} & \mathbf{\Pi}_{12} & \cdots & \mathbf{\Pi}_{1I} \\ \mathbf{\Pi}_{21} & \mathbf{\Pi}_{22} & \cdots & \mathbf{\Pi}_{2I} \\ \vdots & \vdots & \ddots & \vdots \\ \mathbf{\Pi}_{I1} & \mathbf{\Pi}_{I2} & \cdots & \mathbf{\Pi}_{II} \end{bmatrix} +$$

$$+\sum_{k=1}^{K} \begin{bmatrix} \Delta \mathbf{y}_{1,t-k} & \Delta \mathbf{y}_{2,t-k} & \cdots & \Delta \mathbf{y}_{I,t-k} \end{bmatrix} \begin{bmatrix} \mathbf{\Gamma}_{11,k} & \mathbf{\Gamma}_{12,k} & \cdots & \mathbf{\Gamma}_{1I,k} \\ \mathbf{\Gamma}_{21,k} & \mathbf{\Gamma}_{22,k} & \cdots & \mathbf{\Gamma}_{2I,k} \\ \vdots & \vdots & \ddots & \vdots \\ \mathbf{\Gamma}_{I1,k} & \mathbf{\Gamma}_{I2,k} & \cdots & \mathbf{\Gamma}_{II,k} \end{bmatrix} + \qquad (6.75)$$

$$+\begin{bmatrix} \boldsymbol{\varepsilon}_{1t} & \boldsymbol{\varepsilon}_{2t} & \cdots & \boldsymbol{\varepsilon}_{It} \end{bmatrix},$$

where the variance-covariance matrix is unrestricted (see (6.66)). Model (6.75) is equivalent to model (6.48), as no restrictions are imposed on either the matrices of the parameters or the variance-covariance matrix. If the $IP \times IP$-element $\mathbf{\Pi}$ matrix has a reduced rank, then the matrix can be decomposed so that $\mathbf{\Pi} = \mathbf{B}\mathbf{A}^T$, $\mathbf{A}$ and $\mathbf{B}$ are the $IP \times \sum_{i=1}^{I} R_i$ matrices of loadings and cointegration vectors, respectively, and model (6.75) can be written as follows:

$$\Delta \mathbf{y}_t = \mathbf{y}_{t-1}\mathbf{B}\mathbf{A}^T + \sum_{k=1}^{K} \Delta \mathbf{y}_{t-k}\mathbf{\Gamma}_k + \boldsymbol{\varepsilon}_t. \qquad (6.76)$$

Larsson and Lyhagen (1999) propose to decompose the matrix of the long-run multipliers so that the loadings matrix can be given as:

$$\mathbf{A} = \begin{bmatrix} \mathbf{A}_{11} & \mathbf{A}_{12} & \cdots & \mathbf{A}_{1I} \\ \mathbf{A}_{21} & \mathbf{A}_{22} & \cdots & \mathbf{A}_{2I} \\ \vdots & \vdots & \ddots & \vdots \\ \mathbf{A}_{I1} & \mathbf{A}_{I2} & \cdots & \mathbf{A}_{II} \end{bmatrix}, \qquad (6.77)$$

and the cointegrating vectors matrix has the block-diagonal structure:

$$\mathbf{B} = \mathbf{B}_A = \begin{bmatrix} \mathbf{B}_{11} & \mathbf{0} & \cdots & \mathbf{0} \\ \mathbf{0} & \mathbf{B}_{22} & \cdots & \mathbf{0} \\ \vdots & \vdots & \ddots & \vdots \\ \mathbf{0} & \mathbf{0} & \cdots & \mathbf{B}_{II} \end{bmatrix}, \qquad (6.78)$$

where $\mathbf{A}_{ij}$, $j = 1, 2, \ldots, I$, and $\mathbf{B}_{ii}$ are $P \times R_i$ matrices, and the rank of matrix $\mathbf{B}_{ii}$ equals R_i. In the case of the common cointegration rank R, $\mathbf{A}$ and $\mathbf{B}$ are $IP \times IR$ matrices. It should be stressed that the non-diagonal structure of matrix $\mathbf{\Pi}$ results solely from the impact of the error correction terms on the variables in different cross-sections. Therefore, the variables are not allowed to cointegrate

across the cross-sections. The maximum likelihood estimates of the parameters of model (6.76) with matrix $\mathbf{B}_A$ can be calculated using a reduced rank regression with restrictions on the long-run coefficients (see Johansen and Juselius (1992)):
$\mathbf{B} = (\mathbf{H}_1\mathbf{B}_{11}\ \mathbf{H}_2\mathbf{B}_{22} \dots \mathbf{H}_I\mathbf{B}_{II})$,
where $\mathbf{H}_i$ is such an $IP \times P$ matrix that the ith submatrix is a unity matrix and other elements are zeros:
$\mathbf{H}_i = \left[\mathbf{0} \dots \mathbf{0}\, \mathbf{I}_p\, \mathbf{0} \dots \mathbf{0}\right]^T$.

Model (6.76) with the matrix $\mathbf{B}_A$ of cointegrating vectors allows inferring on the common cointegration rank. The asymptotic distribution of the likelihood ratio statistic is convolution of the Dickey-Fuller type distribution and the χ^2 distribution:

$$LR_{IT}\left(\mathcal{H}_0\left(\mathbf{\Pi}_B\right)\middle|\mathcal{H}_1\left(\mathbf{\Pi}_A\right)\right) \mathrm{T}\ \chi^2_{I(I-1)R(P-R)} + \\ +\mathrm{tr}\left(\int_0^1 \left(d\tilde{\mathbf{W}}\right)^T \tilde{\mathbf{F}}\left[\int_0^1 \tilde{\mathbf{F}}^T\tilde{\mathbf{F}}du\right]^{-1} \int_0^1 \tilde{\mathbf{F}}^T\left(d\tilde{\mathbf{W}}\right)\right) \quad \text{as } T \to \infty, \text{ for fixed } I, \tag{6.79}$$

where $\tilde{\mathbf{W}}$ denotes the $I(P-R)$-dimensional Brownian motion and $\tilde{\mathbf{F}}$ is the Brownian motion corrected for a constant or a linear trend. The null hypothesis $\mathcal{H}_0 : \mathbf{B}_{11} = \mathbf{B}_{22} = \dots = \mathbf{B}_{II}$ about the common, exactly identified cointegrating vectors can be verified by means of the likelihood ratio statistic for which the limit distribution is given by:

$$LR_{IT}\left(\mathcal{H}_0\left(\mathbf{\Pi}_C\right)\middle|\mathcal{H}_1\left(\mathbf{\Pi}_B\right)\right) \mathrm{T}\ \chi^2_{(I-1)R(P-R)} \quad \text{as } T \to \infty, \text{ for fixed } I. \tag{6.80}$$

However, model (6.76) with such a restriction cannot be estimated using the reduced rank regression, such as used in the Johansen procedure. Therefore, Larsson and Lyhagen proposed to compute the maximum likelihood estimates by means of the switching algorithm developed by Boswijk (1995).

Even though the decomposition of matrix $\mathbf{\Pi}$ into matrices $\mathbf{A}$ and $\mathbf{B}$ given respectively by (6.77) and (6.78) allows a non-diagonal structure of matrix $\mathbf{\Pi}$, the variables from the distinct cross-sections are not permitted to cointegrate. There are cases, however, when this restriction cannot be met, especially if such relationships are postulated by economic theories. The first solution to this problem was the model proposed by Jacobson et al. (2008), with one variable common for all cross-sections.

Let $\mathbf{y}_t = \left[\mathbf{y}_{1t}\ \mathbf{y}_{2t} \cdots \mathbf{y}_{It}\ y_t^*\right]$ define an $(IP+1)$-element vector of P variables in I cross-sections, where the y_t^* represents a variable common to all cross-sections. Then, the cointegrating vectors matrix can be written as:

$$\mathbf{B} = \begin{bmatrix} \mathbf{B}_{11} & \mathbf{0} & \cdots & \mathbf{0} \\ \mathbf{0} & \mathbf{B}_{22} & \cdots & \mathbf{0} \\ \vdots & \vdots & \ddots & \vdots \\ \mathbf{0} & \mathbf{0} & \cdots & \mathbf{B}_{II} \\ \mathbf{b}_{I+1,1} & \mathbf{b}_{I+1,2} & \cdots & \mathbf{b}_{I+1,I} \end{bmatrix}, \tag{6.81}$$

where $\mathbf{B}_{ii}$ is a $P \times R_i$ full rank matrix and $\mathbf{b}_{I+1,i}$ is a $1 \times R_i$-element vector. Model (6.76) with the $\mathbf{B}$ matrix given by (6.81) is estimated by means of the reduced rank regression with restrictions imposed on the cointegration space. The likelihood ratio statistic of the test for a common cointegration rank (see hypotheses (6.56a-b)) converges asymptotically as follows:

$$LR_{IT}\left(\mathcal{H}_0\left(\mathbf{\Pi}_B\right)\middle|\mathcal{H}_1\left(\mathbf{\Pi}_A\right)\right) \Rightarrow \chi^2_{I(I-1)R(P-R)} + \\ +\mathrm{tr}\left(\int_0^1 \left(d\tilde{\mathbf{W}}\right)^T \tilde{\mathbf{F}}\left[\int_0^1 \tilde{\mathbf{F}}^T \tilde{\mathbf{F}} du\right]^{-1} \int_0^1 \tilde{\mathbf{F}}^T \left(d\tilde{\mathbf{W}}\right)\right) \quad \text{as } T \to \infty, \text{ for fixed } I, \tag{6.82}$$

where $\tilde{\mathbf{W}}$ is the $\left(I\left(P-R\right)+1\right)$-dimensional Brownian motion, and $\tilde{\mathbf{F}}$ is the Brownian motion corrected for a constant or a linear trend.

The null hypothesis $\mathcal{H}_0 : \mathbf{B}_{11} = \mathbf{B}_{22} = \ldots = \mathbf{B}_{II}$ about the common, exactly identified cointegrating vectors can be verified by the likelihood ratio statistic, and the asymptotic distribution is given as:

$$LR_{IT}\left(\mathcal{H}_0\left(\mathbf{\Pi}_C\right)\middle|\mathcal{H}_1\left(\mathbf{\Pi}_B\right)\right) \Rightarrow \chi^2_{(I-1)R(P-R)} \quad \text{as } T \to \infty, \text{ for fixed } I. \tag{6.83}$$

It should be noted that the general model given by (6.76) with the loadings matrix (6.77) and the following cointegration vectors matrix:

$$\mathbf{B} = \begin{bmatrix} \mathbf{B}_{11} & \mathbf{B}_{12} & \cdots & \mathbf{B}_{1I} \\ \mathbf{B}_{21} & \mathbf{B}_{22} & \cdots & \mathbf{B}_{2I} \\ \vdots & \vdots & \ddots & \vdots \\ \mathbf{B}_{I1} & \mathbf{B}_{I2} & \cdots & \mathbf{B}_{II} \end{bmatrix}, \tag{6.84}$$

so that $rk\left[\mathbf{B}_{1i}^T \; \mathbf{B}_{2i}^T \ldots \mathbf{B}_{Ii}^T\right]^T = R_i$ for each i, would be difficult to estimate. The problem is due to the following decomposition of matrix $\mathbf{\Pi}$:

$$\mathbf{\Pi}_{ij} = \sum_{l=1}^{I} \mathbf{B}_{il} \mathbf{A}_{jl}^T, \tag{6.85}$$

which has to be performed instead of the standard decomposition:

$$\mathbf{\Pi}_{ij} = \mathbf{B}_{ii} \mathbf{A}_{ji}^T, \tag{6.86}$$

which is conducted, when the cointegration vectors matrix is given by (6.78).

Applications of the aforementioned models and tests have been sparse so far, because these methods are still not covered by the standard econometric software. It is worth mentioning, though, that Groen and Kleibergen (2003) attempted to verify the monetary approach to the modelling of an exchange rate,

Jacobson at el. (2008) investigated purchasing power parity for France, Germany, Italy and the UK, while Karaman (2008) tested the Fisher hypothesis for the OECD member states.

References

Baltagi B. H., Kao C., 2000, *Nonstationary Panels, Cointegration in Panels and Dynamic Panels: a Survey*, "Advances in Ecnometrics", vol. 15, pp. 7–51.
Banerjee A., 1999, *Panel Data Unit Roots and Cointegration: an Overview*, "Oxford Bulletin of Economics and Statistics", vol. 61, pp. 607–629.
Boswijk H.P., 1995, *Identifiability of Cointegrated Systems*, "Tinbergen Institute Discussion Paper".
Breitung J., Meyer W., 1994, *Testing for Unit Roots in Panel Data: Are Wages on Different Bargaining Levels Cintegrated?*, "Applied Economics", vol. 26, pp. 353–361.
Chen B., McCoskey S., Kao C., 1999, *Estimation and Inference of a Cointegrated Regression in Panel Data: a Monte Carlo Study*, "American Journal of Mathematical and Management Sciences", vol. 19, pp. 75–114.
Choi I., 2001, *Unit Root Tests for Panel Data*, "Journal of International Money and Finance", vol. 20, pp. 249–272.
Engle R.F., Granger C.W.J., 1987, *Co-integration and Error Correction Representation, Estimation, and Testing*, "Econometrica", vol. 55, pp. 251–276.
Entorf H., 1997, *Random walks with drifts: Nonsense regression and spurious fixed-effect estimation*, "Journal of Econometrics", vol. 80, pp. 287–296.
Frankel J.A., Rose A.K., 1996, *A Panel Project on Purchasing Power Parity: Mean Reversion Within and Beetwen Countries*, "Journal of International Economics", vol 40, pp. 209–224.
Groen J.J.J., Kleibergen F., 2003, *Likelihood-Based Cointegration Analysis in Panels of Vector Error Correction Models*, "Journal of Business & Economics Statistics", vol. 21, pp. 295–318.
Hadri K., 2000, *Testing for Stationarity in Heterogeneous Panel Data*, "Econometrics Journal", vol. 3, pp. 148–161.
Hadri K., Larsson R., 2005, *Testing for Stationarity in Heterogeneous Panel Data where the Time Dimension is Finite*, "Econometrics Journal", vol. 8, pp. 55–69.
Im K.S., Pesaran M.H., Shin Y., 2003, *Testing for Unit Roots in Heterogeneous Panels*, "Journal of Econometrics", vol. 115, pp. 53–74.
Jacobson T., Lyhagen J., Larsson R., Nessén M., 2008, *Inflation, Exchange Rates and PPP in a Multivariate Panel Cointegration Model*, "Econometrics Journal", vol. 11, pp. 58–79.

Johansen S., 1988, *Statistical Analysis of Cointegration Vectors*, "Journal of Economic Dynamics and Control", vol. 12, pp. 231–254.

Johansen S., 1996, "Likelihood-Based Inference in Cointegrated Vector Autoregressive Models", Oxford University Press, Oxford.

Johansen S., Juselius K., 1992, *Testing Structural Hypotheses in a multivariate cointegration analysis of the PPP and the UIP for UK*, "Journal of Econometrics", vol. 53, pp. 211–24.

Kao C., 1999, *Spurious Regression and Residual-Based Tests for Cointegration in Panel Data*, "Journal of Econometrics", vol. 90, pp. 1–44.

Kao C., Chiang M.-H., 1997, *On the Estimation and Inference of a Cointegrated Regression in Panel Data*, "Nonstationary Panels, Panels Cointegration and Dynamics Panels", vol. 15, pp. 179–222.

Kao C., Chiang M.-H. Chen B., 1999, *International R&D Spillovers: An Application of Estimation and Inference in Panel Cointegration*, "Oxford Bulletin of Economics and Statistics", vol. 61, pp. 691–709.

Karaman D.D., 2008, *Comparison of Panel Cointegration Tests*, "Economics Bulletin", vol. 3, pp. 1–20.

Kwiatkowski D., Phillips P.C.B., Schmidt P., Shin Y., 1992, *Testing the Null Hypothesis of Stationarity Against the Alternative of a Unit Root*, "Journal of Econometrics", vol. 54, pp. 159–178.

Larsson R., Lyhagen J., 1999, *Likelihood-Based Inference in Multivariate Panel Cointegration Models*, "Stockholm School of Economics Working Paper Series in Economics and Finance", vol. 331.

Larsson R., Lyhagen J., Löthgren M., 2001, *Likelihood-Based Cointegration Tests in Heterogeneous panels*, "Econometrics Journal", vol. 4, pp. 109–142.

Lee G., 2005, *Direct versus indirect international R&D spillovers*, "Information Economics and Policy", vol. 17, pp. 334–348.

Levin A., Lin C.-F., Chu C.-S. J., 2002, *Unit Root Tests in Panel Data: Asymptotic and Finite-Sample Properties*, "Journal of Econometrics", vol. 108, pp. 1–24.

Lütkepohl H., Saikkonen P., Trenkler C., 2001, *Maximum Eigenvalue versus Trace Tests for the Cointegrating Rank of a VAR Process*, "Econometrics Journal", vol. 4, pp. 287–310.

MacDonald R., 1996, *Panel Unit Root Tests and Real Exchange Rates*, "Economics Letters", vol. 50, pp. 7–11.

Maddala G.S., Wu S., 1999, *A Comparative Study of Unit Root Tests with Panel Data and a New Simple Tests*, "Oxford Bulletin of Economics and Statistics", vol. 61, pp. 631–652.

McCoskey S., Kao C., 1998, *A Residual-Based Test of the Null of Cointegration in Panel Data*, "Econometric Reviews", vol. 17, pp. 57–84.

McCoskey S., Kao C., 1999, *Testing the Stability of a Production Function with Urbanization as a Shift Factor: An Application of Non-Stationary Panel Data Techniques*, "Center for Policy Research Working Papers", vol. 5.

Nabeya S., 1999, *Asymptotic moments of some unit root test statistics in the null case*, "Econometric Theory", vol. 15, pp. 139–149.

Pedroni P., 2000, *Fully Modified OLS for Heterogeneous Cointegrated Panels*, "Nonstationary Panels, Panel Cointegration and Dynamic Panels", vol. 15, pp. 93–130.

Pelgrin F., Schich S., 2002, *Panel Cointegration Analysis of the Finance-Investment Link in OECD Countries*, "Documents de Travail de l'OFCE", vol. 2002-02, pp. 1–24.

Phillips P.C.B., Moon H.R., 1999, *Linear Regression Limit Theory for Nonstationary Panel Data*, "Econometrica", vol. 67, pp. 1057–1111.

Phillips P.C.B., Moon H.R., 2000, *Nonstationary Panel Data Analysis: an Overview of Some Recent Developments*, "Econometrics Reviews", vol. 19, pp. 263–286.

Phillips P.C.B., Perron P., 1988, *Testing for a Unit Root in Time Series Regression*, "Biometrica", vol. 75, pp. 335–346.

Quah D., 1990, *International Patterns of Growth: I. Persistence in Cross-Country Disparites*, "MIT Working Paper".

Quah D., Sargent T.J., 1992, *A Dynamic Index Model for Large Cross-Country Disparities*, mimeo.

Said S.D., Dickey D.A., 1984, *Testing for Unit Roots in Autoregressive-Moving Average Models of Unknown Order*, "Biometrica", vol. 71, pp. 599–607.

Spanos A., 1999, "Probability Theory and Statistical Inference. Econometric Modeling with Observational Data", Cambridge University Press, Cambridge.

Final remarks

This monograph shows the results of an attempt at summarizing the vast contributions to the world literature that have been made mainly in the last 25 years. The authors wished to provide a solid theoretical background, while presenting a systematic review of methods and results related to empirical research conducted in that period. This book provides a stepping-stone for research efforts aimed at introducing suitable extensions to the long-term macroeconometric models. These extensions should include, above all, the analyses of the knowledge capital impacts, thus enabling the construction of long-term forecasts and scenarios based on model simulations. It seems, however, that many issues addressed in this monograph have not been satisfactorily resolved and need further research.

The literature of the subject frequently raises the question, whether an increase in human capital and/or R&D expenditures makes GDP grow, or whether the increase relationship is true, according to which a GDP inverse creates conducive conditions for boosting human capital formation and increasing R&D expenditures. This dilemma seems to be spurious, because in fact both these relationships take place, however not simultaneously and under certain conditions. The research results reported in chapters 4 and 5 indicate quite distinctly that expanding cumulated domestic and foreign R&D expenditures, as well as increasing human capital, do make potential GDP grow with some delay. The intensity of this relationship is illustrated by the positive values of the potential GDP elasticities with respect to the indicated factors of growth. In contrast, the inverse relationship takes effect with long lags. Therefore, an increase in the potential GDP and an increase in the effective GDP (representing the realization of final demand) are not equal in the short term and medium term. The effective GDP expands at a rate similar to the growth rate of the potential GDP, provided that the rate of utilization of the potential output does not change. This expansion is faster (slower), if the degree of capacity utilization goes up (down). To circumvent this difficulty in investigations based on the cross-section time series for long periods, the information on the average rates of growth for 5 or 10-year-periods spanning one or two business cycles is used.

As regards research taking advantage of country-specific time series, attempts are being made to estimate the particular rates of utilization of productive capacities. In general, a macroeconomic model should contain the sector generating final demand in addition to the supply side. It is then possible to establish the relationship between the effective GDP and the demand for investments in physical capital and knowledge capital. It takes time for an increase in fixed capital and many years are needed to improve real human capital. Therefore, these processes can be analysed only after the macroeconomic models have been appropriately extended. A sequence of equations referring to the system of education has to go on stage, to account for the educational process ending when students leaving schools enter labour force thus increasing its quality. It leads to a relevant increase in human capital per employee, after allowing for its depreciation and changes in the rate of unemployment. The outlined feedbacks are not simultaneous, but take effect with long lags. Our earlier analyses have demonstrated that after the relationships are taken into account, it is possible to calculate sustained multipliers revealing the overall long-term effects of policies aimed at improving the absorption of foreign knowledge capital, and affecting broadly understood education.

The results presented in this monograph indicate the necessity of expanding the research spectrum. This first of all applies to efficiency analyses of technological progress embodied in fixed capital and human capital. As regards the analyses of total factor productivity growth, it seems necessary to allow for the impact of computerization on technological processes and management by decomposing fixed capital into computers, software, telecommunication and the other items. The impacts of foreign knowledge should be elaborated in more detail. Beside trying to describe the indirect channels of transfer of the cumulated R&D expenditures via the import of investment and/or intermediate commodities, an attempt should be made to estimate the impact of the direct channels of foreign knowledge transfer. The relevant aspects of this estimation include the influence of the use of patents and licenses, the role of technological proximity and the extension of telecommunication lines, including the use of the Internet.

The concept of human capital should be treated more broadly. It should encompass changes not only in the composition of employees by their level of education, but also by sex, age and industry. It is necessary to define new, extended measures of human capital and calculate their values. The measures need to address the impacts of work experience (learning by doing), post-graduate education, literacy, Internet skills, as well as health conditions.

The application of the above suggestions goes beyond the national economy treated as a whole. It seems necessary to disaggregate the totals by sector, as this is crucial to the estimates of long-term changes in the employment structure. Further disaggregation of labour force by sex, age and the sections of economy seems necessary. This approach is especially important with respect to studying the long-run changes in the rates of unemployment.

An important contribution to the development of the analytical framework would be the construction of satellite models of the research and education sections. This would help to better understand the sources of growth of R&D and educational expenditures.

The implementation of the suggested programme enabling further development of modelling the impact of knowledge capital will involve an appropriate extension of the databases to include new variables and new data covering the relevant cross-sections.

Author Index

FSC
www.fsc.org
MIX
Papier | Fördert gute Waldnutzung
FSC® C083411

Zeitfracht Medien GmbH
Ferdinand-Jühlke-Straße 7
99095 Erfurt, Deutschland
produktsicherheit@kolibri360.de